T0227413

Maurício Ehrlich | Leonardo Becker

Reinforced Soil Walls and Slopes

design and construction

Maurício Ehrlich | Leonardo Becker

Reinforced Soil
Walls and Slopes
design and construction

CRC Press
Taylor & Francis Group
Boca Raton London New York

CRC Press is an imprint of the
Taylor & Francis Group, an **informa** business

CRC Press
Taylor & Francis Group
6000 Broken Sound Parkway NW, Suite 300
Boca Raton, FL 33487-2742

© 2010 by Taylor & Francis Group, LLC
CRC Press is an imprint of Taylor & Francis Group, an Informa business

No claim to original U.S. Government works

This book contains information obtained from authentic and highly regarded sources. Reason-able efforts have been made to publish reliable data and information, but the author and publisher cannot assume responsibility for the validity of all materials or the consequences of their use. The authors and publishers have attempted to trace the copyright holders of all material reproduced in this publication and apologize to copyright holders if permission to publish in this form has not been obtained. If any copyright material has not been acknowledged please write and let us know so we may rectify in any future reprint.

Except as permitted under U.S. Copyright Law, no part of this book may be reprinted, reproduced, transmitted, or utilized in any form by any electronic, mechanical, or other means, now known or hereafter invented, including photocopying, microfilming, and recording, or in any information storage or retrieval system, without written permission from the publishers.

For permission to photocopy or use material electronically from this work, please access www.copyright.com (http://www.copyright.com/) or contact the Copyright Clearance Center, Inc. (CCC), 222 Rosewood Drive, Danvers, MA 01923, 978-750-8400. CCC is a not-for-profit organiza-tion that provides licenses and registration for a variety of users. For organizations that have been granted a photocopy license by the CCC, a separate system of payment has been arranged.

Trademark Notice: Product or corporate names may be trademarks or registered trademarks, and are used only for identification and explanation without intent to infringe.

Visit the Taylor & Francis Web site at
http://www.taylorandfrancis.com

and the CRC Press Web site at
http://www.crcpress.com

To Ana Paula and João Gabriel
(*L. Becker*)

To Dorise and Fabio, who are always present
(*M. Ehrlich*)

About the Authors

Leonardo De Bona Becker is a Professor at the Federal University of Rio de Janeiro. He earned titles in Civil Engineering and MS in Geotechnical Engineering from Universidade Federal do Rio Grande do Sul, respectively in 1999 and 2001, and has a PhD in Geotechnical Engineering from the Catholic University of Rio de Janeiro in 2006. Published technical papers in national and international congresses, and was recognized by the Brazilian Association of Soil Mechanics and Geotechnical Engineering with the Icarahy da Silveira award, in 2002. His experience includes construction, design and consulting related to retaining structures, earth dams and port structures.

Professor Dr. Maurício Ehrlich, D.Sc. COPPE / UFRJ and Postdoctoral at Berkeley, University of California, is a professor at the Federal University of Rio de Janeiro (UFRJ) and a researcher supported by the Brazilian Research Council (CNPq) and by the Rio de Janeiro State Funding Agency (FAPERJ). Ex-Chairman of the Civil Engineering Department at the Graduate School of Engineering (COPPE) and current President of the Brazilian Association of Geosynthetics (IGS Brazil). He is a visiting professor at universities in England, the United States and Germany. Dedicated to teaching, research and consulting in the areas of geotechnical retaining structures, excavations, slopes and environmental geotechnics. His scientific production involves more than 180 technical papers published both nationally and internationally. He supervised 18 Ph.D. theses and 29 Masters Degree dissertations. His work is recognized in Brazil and abroad and has received the Norman Medal from the American Society of Civil Engineering (ASCE) and the Terzaghi Award, from the Brazilian Association of Soil Mechanics and Geotechnical Engineering (ABMS).

Contents

Notation, 11

Presentation, 15

Preface, 17

Chapter 1 — Introduction, 19

Chapter 2 — Soil - Reinforcement Interaction, 23

2.1 CONCEPTS OF REINFORCED SOIL 23

2.2 CHARACTERISTICS OF GEOSYNTHETICS FOR REINFORCEMENT 24

2.3 SOIL BEHAVIOR 30

2.4 INFLUENCE FACTORS ON REINFORCEMENT TENSION 35

Chapter 3 — Analysis and design, 47

3.1 INTRODUCTION 47

3.2 EXTERNAL STABILITY 47

3.3 INTERNAL STABILITY 49

3.4 DETERMINATION OF MAXIMUM REINFORCEMENT TENSION 51

3.5 DETERMINATION OF THE LOCATION OF REINFORCEMENT MAXIMUM TENSION

　　　POINT 61

3.6 PULLOUT ANALYSIS 63

3.7 REINFORCEMENT AND FACING CONNECTIONS 64

3.8 DESIGN RESISTANCE OF REINFORCEMENTS 65

3.9 STRESS INCREMENTS DUE TO SURCHARGE LOADINGS 66

3.10 LATERAL DISPLACEMENT AND STRAIN PREDICTION 71

3.11 CASE STUDIES 75

Chapter 4 — Constructive Aspects, 77

4.1 TYPE OF REINFORCEMENT 77

4.2 TYPE OF FACING AND CONSTRUCTION PROCEDURE 78

4.3 ARRANGEMENT OF REINFORCEMENTS 83

4.4 DRAINAGE SYSTEM 84

4.5 BACKFILL MATERIAL 85

4.6 QUALITY CONTROL AND CONSTRUCTIVE TOLERANCES 88

4.7 GUARDRAILS 88

Chapter 5 — Design Example, 89

5.1 SOIL AND WALL CHARACTERISTICS 89

5.2 EXTERNAL STABILITY ANALYSIS 90

5.3 INTERNAL STABILITY ANALYSIS 93

5.4 COMPARISON OF METHODS FOR REINFORCEMENT TENSION DETERMINATION 98

5.5 WALL FINAL ARRANGEMENT 99

Appendix – Design charts, 101

References, 109

Notation

$(\sigma_1 - \sigma_3)$	deviatory stress
$(\sigma_1 - \sigma_3)_{ult}$	ultimate deviatory stress
$(\sigma'_h)_{ave}$	average maximum horizontal earth stress
a	adhesion of soil-geosynthetic interface
B	width of the plate
B'	reduce equivalent width of wall foundation
X	dimensionless parameter obtained from the Ehrlich and Mitchell chart
c'	effetive cohesion intercept of the soil
CR	coupling efficiency between face and reinforcement coefficient
D	wall embedment depth
D_{15}	diameter at which 15% of the soil is finer
D_{85}	diameter at which 85% of the soil is finer
E	Earth pressure resultant
E_i	initial tangent modulus
E_{ur}	unloading and reloading tangent modulus
ϕ	soil friction angle
f_f	creep coefficient
$F*$	coefficent of pullout resistance
f_a	adhesion ratio
f_a	biochemical degradation coefficent
$f_{ci,qi,\gamma i}$	load eccentricity and inclination coefficients
ϕ_{cv}	soil friction angle at critical state
f_d	resistance reduction factor for mechanical damage during installation factor
FS	safety factor
FS_a	pullout safety factor
FS_g	overall safety factor of the structure
H	height of reinforced soil structure
J_r	tensile stiffness modulus of reinforcement
K	modulus number (hyperlolic stress-strain curve model)
K	earth pressure coefficient
K_a	active earth pressure coefficient
K_c	earth pressure coefficient during compaction
K_0	at rest earth pressure coefficient

K_r	residual earth pressure coefficient at end of construction
K_u	modulus number for unloading (hyperlolic stress-strain curve model)
K_{ur}	unloading and reloading modulus number (hyperlolic stress-strain curve model)
$K_{\Delta 2}$	at-rest decremental lateral earth presure coefficient for unloading
L	roller drum length
L_a	length of reinforcement in the active zone
L_e	length of reinforcement beyond the failure surface
L_0	length of reinforcement overlap (geotextiles)
L_r	total length of reinforcement
n	modulus expoent (hyperbolic stress strain curve model)
$N_{c,q,\gamma}$	soil bearing capacity factors
OCR	overconsolidation ratio
P_a	atmospheric pressure
P_r	pullout resistance per transverse length unit of reinforcement
$P_{r,0}$	admissible resistance of reinforcement connection at face
Q	maximum vertical operating roller drum force (mass times the dynamic amplification factor)
q	surcharge loading
$Q_{1,2}$	external loads resultant at the top of the wedges (wedges method)
q_{ult}	ultimate bearing capacity
$R_{1,2}$	reactions of stable soil (wedges method)
R_f	failure ratio (hyperbolic stress-strain curve model)
R_h	horizontal load resultant
R_v	vertical load resultant
S_i	relative soil-reinforcement stiffness index
S_v	vertical spacing of reinforcements
t_d	time reference for the project
$T(z)$	mobilized tension in the reinforcement at (z) depth, measured from the crest
$T_1(x)$	distribution function of mobilized tension in the reinforcement along the active zone
$T_2(x)$	distribution function of mobilized tension in the reinforcement along the resistance zone
T_d	design geosynthetic tensile load
$T_k(t)$	long term reinforcement resistance at the reference time of the project
T_{max}	maximum reinforcement tension
T_0	tensile force in the reinforcement at facing
T_r	limit tensile force of a geosynthetic
u	lateral displacement
u_{max}	maximum lateral displacement

$W_{1,2}$	weights of each wedge (wedges method)
z	depth
z_c	compaction influence depth
z_{eq}	equivalent depth including aplication of external vertical load
α	scale factor
β	relative soil-reinforcement extensibility
γ	unit weight of the soil
δ	mobilized friction between wall (reinforced zone) and field (unreinforced zone), or soil-geosynthetic interface friction angle
ΔT_{max}	increment of maximum tension in the reinforcement
$\Delta \sigma_z$	increment of vertical stress due to external load application at T_{max} location
ε_r	maximum elongation of the geosynthetic at failure in the test of wide-width strip method
$\varepsilon_1(x)$	strain distribution function in the active zone
$\varepsilon_2(x)$	strain distribution function in the resistant zone
ν_0	Poisson's ratio at rest
ν_{un}	Poisson's ratio for unloading at rest condition
σ'_h	effective horizontal stress
$\sigma'_{sx,c}$	horizontal stress due to compaction
ΣT_l	summation of required tension in the reinforcements intercepted by the L "i" segment (wedge method)
$\Sigma T_{d,i}$	summation of the design resistance of the reinforcements of the wedge "i" (wedge method)
$\sigma'_{xp,i}$	peak induced horizontal soil stress due to compaction for no horizontal deformation on the reinforcement direction
σ'_z	effective vertical stress
$\sigma'_{z,b}$	stresses acting on the base of the wall
σ'_{zc}	maximum past effective vertical stress including compaction
$\sigma'_{zc,i}$	maximum effective vertical stress induced by compaction
ω	inclination of the structure's face

PRESENTATION

Although Engineering materials are relatively new, with nearly 50 years of existence in the world market, geosynthetics have become compulsory in almost all infrastructure, drainage applications, waterproofing, paving, erosion control and soil reinforcement works. In Brazil, walls and reinforced soil slopes with geosynthetics have been built since the 1980s; however, they gained a strong momentum in the last decade, with the introduction of geosynthetics and better performing construction systems. Presenting a book written by the engineers Maurício Ehrlich and Leonardo Becker on walls and slopes reinforced with geosynthetics is a very rewarding mission, because these are people who deserve all my respect and admiration. In the professional sphere, they have a consistent career in the Geotechnical Engineering academic area of teaching and research, as they also maintain a strong link with the practice of engineering. I believe they are currently listed among the greatest Brazilian and world specialists in the area of geosynthetically-reinforced soils.

Mauricio Ehrlich does not require further presentations. A Professor at COPPE / UFRJ for many years, he is nationally and internationally recognized for his technical contributions to Geotechnical Engineering. I met Professor Ehrlich in the early 1990s, when he was taking his post-doctorate at the University of Berkeley, California, under Professor Mitchell's coordination. At the time, Ehrlich was working on the conceptualization and validation of a new specification method for retaining structures on reinforced soil, a method that considered the working conditions of these structures. The limit equilibrium design methods used until then considered only the peak resistance of materials (soil and reinforcement), without taking into account each one's stiffness, implying that the breakdown of all materials occurred at the same time. The rational method proposed by Ehrlich and Mitchell went on to consider the reinforcement and soil stiffness properties and the effect of backfill compaction as input for the design. This method was published in the geotechnical journal of the ASCE (American Society of Civil Engineers) in 1995, and the article was awarded the Norman Medal, a prize awarded to the best paper published that year.

Professor Leonardo Becker is certainly one of the most brilliant geotechnical engineers of his generation, and his master's thesis received an ABMS (Associação Brasileira de Mecânica dos Solos) award. I had the opportunity to interact extensively with him during his doctorate when he followed up and monitored the execution of a retaining structure of great responsibility, built with fine compacted soil and reinforced with a PVA geogrid to raise the elevation of a containment dam for chemical residues. He also studied the behavior of geogrid pullout in the soil of the dam.

In this period, he overcame all the adversity inherent in a field research and managed to complete his work satisfactorily. He is currently a professor and head of the laboratory of Soil Mechanics at UFRJ.

This book is intended for those who wish to design and build retaining walls and steep slopes on geosynthetically-reinforced soils. The rational design method proposed by Ehrlich and Mitchell and its applications are thoroughly detailed here. Tables with data typical of Brazilian soils and design charts are also presented, as to allow immediate use of this method. Special attention is given to the practical aspect and various issues of different construction systems are discussed. By sponsoring the edition of this book, Huesker hopes to contribute, in a technical and responsible manner, to the development of this application in Brazil. This is the first book of the "Huesker Collection - Engineering with Geosynthetics", launched in partnership with Oficina de Textos in order to disseminate technical knowledge about geosynthetics and their applications, with special attention to the particular conditions of Brazilian soils.

Engineer Flavio Montez
Director Huesker Ltd. (Brazil),
a subsidiary of Huesker GmbH (Germany)

PREFACE

The use of reinforcements represents a revolution to the concept generally used in geotechnical projects. Not that this is indeed a new practice, as fibrous elements associated with soils have been in use for thousands of years. Incas, Babylonians and Chinese adopted soil, bamboo and straw in their constructions. In addition, birds, such as the Red Ovenbird, and beavers use similar procedures.

However, it is understood that the rational use of the technique began in the 1960s, with Henri Vidal, who patented and developed a methodology for a system design that went on to be called "Terre Armée". Externally, a reinforced soil wall is no different from a conventional structure. The particular issue is internal stability. Much like in the structures of reinforced concrete, reinforcement allows for better mechanical performance of the soil. Slopes and landfill heights that would otherwise be impossible, become feasible.

This book aims to bring together, in a didactic manner, knowledge that could guide the design of reinforced soil walls and slopes. The book also aims to elucidate the basic mechanisms associated with soil-reinforcement interaction. The aim is to adequately present the concepts for the practical needs of engineers involved in design, construction and supervision of reinforced soil structures, without neglecting the theoretical rigor.

The traditional analysis philosophy is generally based on limit equilibrium methods or strongly empirical procedures, and it restricts the use of fine soil. Throughout this book, a more rational approach to the problem is presented, discussing the underpinning concepts of stress-strain compatibility between soil and reinforcements, without neglecting the good performance of fine-grained tropical soil, which is abundant in Brazil. It is a great challenge for Geotechnics to advance in modeling actual behavior, enabling the development of safer and more economic projects.

INTRODUCTION 1

Retaining structures are indispensable elements of a great variety of engineering projects such as bridges, roads, piers, docks, railways, buildings in general, power plants, dams etc. Its function is basically to withstand earth pressures, avoiding landslide, allowing the use of the space in front of the slope and on its upper part.

For many years, traditional reinforced or cyclopean concrete retaining structures, whether anchored or not, were the only available option. However, with increased soil heights to contain or in cases of less competent foundation soils, the cost of these structures increases considerably.

To overcome these difficulties, various types of retaining structures in reinforced soils were developed, which due to its advantages, gradually replaced traditional structures, in most cases. These structures are, in a final analysis, gravity walls where the construction material of the structure consists of soil associated with tensile resistant inclusions.

According to Mitchell and Villet (1987), there are cases of retaining structures built with soil reinforced by vegetable fibers long before the emergence of reinforced concrete. There are examples of applications of this association by the Mesopotamians about 5 thousand years ago, such as the Ziggurat of Ur, in modern Iraq. Ziggurats were temples built by layers of mud reinforced with reeds and branches, whose heights reached 50 meters or more. The facing was composed of mud bricks dried in the sun, held together by asphalt grout/mortar. In China, branches of trees were used for reinforcement of earth dams for at least a millennium, and the famous Great Wall of China has stretches of reinforced soil. Figure 1.1 shows an artistic representation of the Ziggurat of Ur.

Despite this history, the modern structural concept of reinforced soil for retaining structures was developed only during the 1960s, in France, by

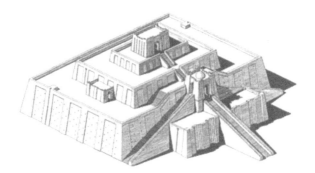

Figure 1.1 *Ziggurat of Ur: perspective*

Henry Vidal (Vidal, 1969), which, in the 1960's and 1970's, disseminated a technique with metallic inclusions called 'reinforced earth'. This is a registered trademark process, but standardized by ABNT (NBR 9286/86), which consists of a reinforced soil wall with galvanized steel strips and cruciform concrete facing.

In the early 1970s, reinforced soil systems with metal inlays were disseminated around the world. Around that same time, the first applications of walls reinforced with polymeric fibers (geosynthetics) came about. Geosynthetics are one of the newest groups of construction materials, commonly used in works of various sizes, especially in heavy construction. The term derives from the junction of "geo", referring to the earth, and "synthetic", referring to the polymeric materials used in their manufacture.

The geotextile was the first type of geosynthetic systematically used in geotechnical engineering, having been used as of the 1950s, in the United States as drainage, separation or erosion control elements. Its use in Europe began in the 1960's and in Brazil, it started in the 1970's. Although there are records of empirical applications of geotextiles in the stabilization of slopes in Brazil dating back to the 1970's, a rational design of a reinforced soil wall came to be known only in 1986 (Carvalho, Pedrosa, Wolle, 1986).

Based on the philosophy of reinforced soil and the rapid development of the petrochemical industry, in the following decades several synthetic materials of high tensile strength and capable of reinforcing the soils began to emerge. The development of geosynthetics was fast. Currently, geogrids are the most commonly used geosynthetic for soil reinforcement. Such material is usually a cheaper alternative and easier to employ in comparison to the existing traditional solutions. Therefore, in a matter of four decades, it went from having the status of incipient technology to cutting-edge technology, with wide acceptance and growing demand. In the early 1990's, this industry already moved billions of dollars (Koerner, 1990). The technique of reinforced soil is used in projects of roads, ports, canals, mining, slope containments and urbanization, among others.

According to Elias, Christopher and Berg (2001), the retaining structures in reinforced soils are cost-effective solutions, which can provide high tolerance for foundation settlements, constructive ease and reduced execution time. The advantage of not requiring skilled labor can also be added. In addition, it also allows for stable soil slopes, in vertical positions, and a good aesthetic finish, if the appropriate facing system is used, such as segmental blocks or slope re-vegetation.

Figure 1.2 presents a cost comparison for some types of retaining structures.

A typical arrangement of a reinforced soil wall is represented in Fig 1.3, which highlights the constituting elements.

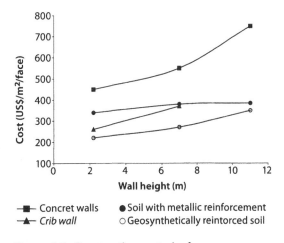

Figure 1.2 *Construction costs, by face area, according to the height of the wall, for various retaining solutions (Elias, Christopher; Berg, 2001)*

Figure 1.4 shows the appearance of a reinforced soil wall built in 2004 for the lowering of a railway line that crosses the city of Maringa, in Parana. The maximum height of the wall is 9 meters. A mature residual soil available on site, classified as sandy clay was used, with approximately 40% fines (sieve # 200). The facing of the wall was executed with Terrae-W® concrete blocks and the reinforcements are Fortrac® polyester geogrids.

Since its first applications, the technique has been studied by several researchers in order to understand the behavior mechanisms and

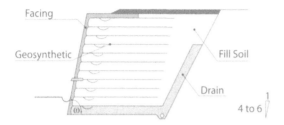

Facing

Geosynthetic

Fill Soil

Drain

1
4 to 6

Figure 1.3 *Cross section of a typical reinforced soil wall with geosynthetics (Ehrlich, Azambuja, 2003)*

develop specification methods. As a result, today there are several design methods for reinforced soil structures, based on empirical, limit equilibrium or working stress conditions considerations.

The methods lead to significantly different results due to their differing assumptions.

The aim of this book is to organize and present the most relevant aspects of reinforced soil retaining structures design and construction (RSRS) with geosynthetics, based on studies conducted by different authors, with emphasis on design methods and recommendations found in Ehrlich and Mitchell (1994), Ehrlich (1999), Dantas and Ehrlich (2000a) and Ehrlich and Azambuja (2003).

Figure 1.4 *Appearance of a reinforced soil wall with segmental block facing (Brugger, 2007)*

SOIL - REINFORCEMENT INTERACTION 2

2.1 CONCEPTS OF REINFORCED SOIL

The association of soil and reinforcements leads the composite material to have better mechanical properties. The soil, when properly compacted, in general, has good resistance to compression and shear. The tensile strength, however, is low or zero. Similarly, to what happens in reinforced concrete, the inclusion of reinforcements makes up for this deficiency. For example, because they have zero tensile strength, clean sands have maximum slopes limited to the rest angle of these materials, which is about 40° (Fig. 2.1A). These same sands, if reinforced, would allow for vertical slopes (Figure 2.1B). Reinforcements act in a similar fashion to a "seam", commiserating the region where the soil would be potentially unstable to the stable region and preventing the collapse of the embankment.

Great differences in level can be achieved, according to the quantity and the resistance of the reinforcements used. The length and amount of reinforcements are determined by analysis of external and internal stability.

In general, the length of geosynthetics varies between 60% and 80% of the height of the wall.

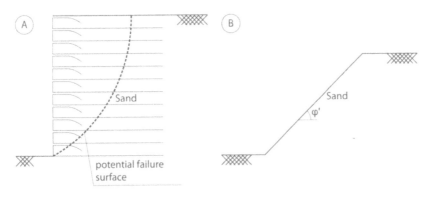

Figure 2.1 *Influence of reinforcement on slope stability: (A) reinforced sand; (B) sand slope (adapted from Ehrlich and Azambuja, 2003)*

2.2 Characteristics of Geosynthetics for Reinforcement

Various types of reinforcements for RSRS can be used: metallic reinforcements, geogrids, woven or nonwoven geotextiles. A wide variety of types of geosynthetic reinforcements have been developed in recent decades for soil reinforcement purposes, divided into two main categories: geotextiles and geogrids. Geotextiles are the most versatile geosynthetic. In addition to reinforcing the soil, they can be used for draining, filtration and separation. Geotextiles can be divided into woven or nonwoven groups, according to their manufacturing process.

Geogrids are specially produced for soil reinforcement. They are very open flat plastic structures, in a grid format, hence the name geogrid. Details of the different types of reinforcements are presented below.

2.2.1 Metallic reinforcements

As noted above, the pioneer reinforcements of the modern era were the steel strips used in the Reinforced Earth system. The dimensions of the strips vary according to the demands of each work. The thickness should include an additional value (sacrificial thickness) to compensate for losses due to corrosion. According to the useful life of the work and the degree of aggressiveness of the environment, the sacrificial thickness, defined as recommended by ABNT (NBR 9286/86), may cause the work to be more expensive. Gabion mesh can also be used as reinforcement with the same restrictions as to corrosion.

2.2.2 Woven geotextiles

The woven geotextiles are manufactured using conventional weaving processes. Basic polymers are melted and forced to pass through an extruder, which grants them the form of filaments, which take the shape of the holes of the extruder, and may be cylindrical or laminetes. Once cooled, the filaments are woven in two directions. The longitudinal direction of the machine is called warp and the cross section is called weft. The intertwining of the fibers may be in the form of single or multiple filaments and laminetes.

2.2.3 Nonwoven geotextiles

They are basically made with the same types of filaments as the previous group, cut into pieces or continuous. However, instead of the weaving processes, the filaments are launched randomly on a conveyor belt, on top of each other. The weight of the product depends on the release rate of the filaments.

The bonding of the filaments is consolidated by means of thermal processes, resin tapping with chemicals or needle punched (small needles with fins weave the filaments).

After the bonding phase of the filaments, the geotextile can be further pressed into rolls of smaller diameter for storage, which are easiest to transport. Nonwoven geotextiles have great structural complexity, in addition to presenting more isotropic physical and mechanical characteristics than woven geotextiles.

Figure 2.2 shows blown up photos of woven and nonwoven geotextile structures.

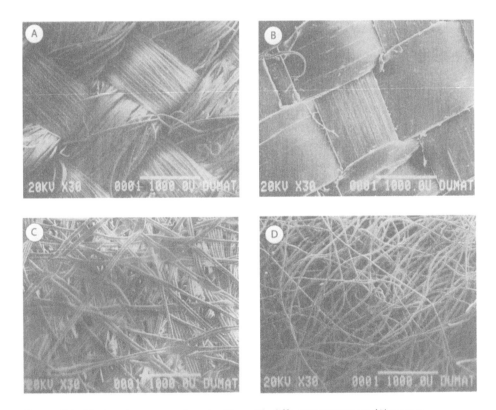

Figure 2.2 *Photo microscopy of geotextiles with different structures: (A) woven multifilament, (B) woven monofilament, (C) non woven needle punched, (D) nonwoven thermo bonded (Koerner, 1998)*

2.2.4 **Geogrids**

Like geotextiles, the geogrids are supplied in rolls of specific width and length. The two main types are: one-way, when they have high tensile strength in only one direction, and two-way, when they have high tensile strength in two orthogonal directions.

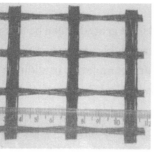

The commonly used polymers for geogrid manufacture are high-density polyethylene (HDPE), polyester (PET) and polyvinyl alcohol (PVA), and they are characterized by low deformability and high tensile strength. Figure 2.3 shows the appearance of a piece of flexible geogrid, made from PVA threads with a polymer protection coating.

Figure 2.3 *Photograph of a PVA geogrid*

2.2.5 **Types of polymers**

The vast majority of geosynthetics is made up of synthetic polymers. The molecular structure of a polymer can be compared to the gathering of many molecule chains that are repeated, called monomers. The most commonly used polymers in geosynthetic manufacturing are polyester (PET), polypropylene (PP), polyethylene (PE) and polyvinyl alcohol (PVA). They consist of long chains of molecules arranged in crystalline regions (aligned chains) and amorphous regions (randomly intertwined chains), obtained by chemical polymerization.

2.2.6 **Creep and deformability**

The non-confined tensile strength is determined in the test of wide-width strip method, with a sample 200mm wide by 100mm long, set between two clamps and pulled until failure. The load applied at the time of rupture divided by the initial width of the sample body is the tensile strength, expressed in kN/m.

The test is standardized by ABNT (NBR 12824/93).

Monitoring of loads and strains is performed during the wide-width test, on which the distributed load-strain curve of the geosynthetics can be based. The shape of the curves varies greatly depending on the type of geosynthetic, its polymer and structure. Figure 2.4 shows some typical

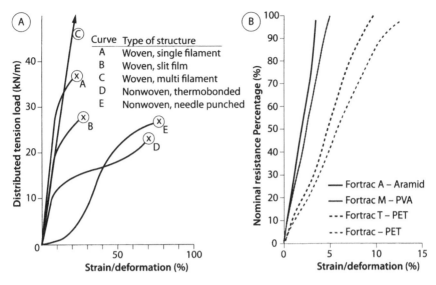

Figure 2.4 *Behavior of distributed load-strains of various types of geosynthetics: (A) geotextiles (Koerner, 1990), (B) geogrids (Huesker)*

examples. It is observed that woven geotextiles have much greater stiffness than non-woven, due to loose bonding between the nonwoven filaments. The stiffness of geogrids, in turn, depends mainly on the polymer used in its manufacture.

In general, the curves are not linear and thus different types of stiffness can be calculated. It is customary to use the initial tangent stiffness, which is the inclination of the initial stress-strain curve and the secant stiffness to the origin, which is the inclination of a straight line connecting the origin to a point on the curve - for example, 2% strain/deformability.

The stiffness is expressed in kN/m and can be defined as follows:

$$J_r = \frac{T_r}{\varepsilon_r}$$

where J_r is the reinforcement tensile strength stiffness module; T_r is the tensile strength per width unit in the wide-width strip test; and ε_r is the corresponding strain of the sample in the wide-width strip test.

For many geogrids, it is unusual for manufacturers to provide secant stiffness for 2% strain or at failure stress.

Soil confinement generally increases the tensile strength and stiffness of nonwoven geotextiles. Because these geosynthetics have filaments that are arranged in a crooked manner, not aligned with the loading, the confinement increases the internal friction between the filaments, restricting slides and reorientation. The soil grain entry into the geotextile structure also helps to restrict the movement of the filaments, increasing stiffness and final resistance.

In geogrids, due to its much simpler structure, the confinement effect on resistance and deformability properties is negligible.

Creep is a common property of polymers. The term creep is employed for the phenomena of deformation of materials over time under constant loading solicitation. Another rheological phenomenon associated to creep is stress relaxation, which occurs when a material is kept at constant strain over time and the internal stresses are reduced.

If, after a while, the material experiences failure under a tension load lower than its short-term tensile strength, which was determined by wide-width strip testing, this phenomenon is called creep failure. Creep tests are usually conducted in laboratory with controlled temperature and without confinement. There is evidence that confinement changes the creep properties of non-woven geotextiles (McGown; Andrawes, Kabir, 1982; Becker, 2001). For woven geotextiles and, in particular for geogrids, the effect of confinement for creep is negligible.

Most creep studies are based on non-confined trials, recently standardized by ABNT (NBR 15226/05). The sample is held vertically between two clamps and weights are attached to the lower clamp for stress/tension. The weights ensure constant load during the desired period of time. The strains of the sample are recorded at regular intervals, and the temperature and humidity of the enclosure should be kept constant. The load is expressed in kN/m, by dividing the applied weight by the original width of the geotextile. Tests can be conducted with loads of 10%, 20%, 30%, 40% and 60% of tensile strength. Strains are calculated based on the initial length of the sample. The test results can be expressed in many ways, and the most common are presented in Figure 2.5.

In general, the designer should prevent reinforcements from reaching the ultimate failure state or from suffering excessive strains that affect the appearance or use of the structure. Watts, Brady and Greene (1998) suggest that limiting the strain imposed on the reinforcement is sufficient to meet both criteria.

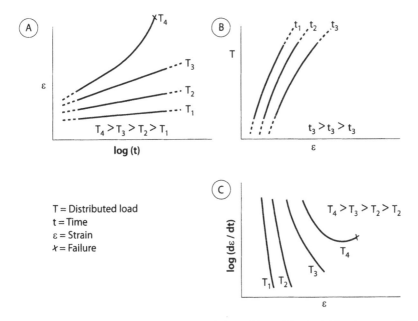

Figure 2.5 *Idealized creep graphs type: (A) semi-logarithmic, (B) isochrones, (C) Sherby-Dorn (Cazzuffi et al., 1997)*

In practice, two approaches can be used to solve the problem:

▶ Reduce the short-term tensile strength by dividing it by a reduction factor that takes into account the type of polymer and considers the useful lifetime of the work and acceptable strains in an implicit and empirical manner. This method is used when there are no available results for creep tests.

▶ Establish a maximum strain value allowed for the reinforcement and determine the load corresponding to the useful lifetime of the work in the isochronous graph.

In both cases, the burden will be further diminished by other reduction factors in order to consider installation damage, variability of the material and chemical and biological degradation.

The strain/deformation value to be used for the reinforcement depends on the work's responsibility, the properties of the soil and tolerable strains for the structure.

It is worth mentioning that the creep behavior of a geosynthetic depends on several factors, such as the nature of the polymer, the structure of the geosynthetic, loading intensity and temperature.

The most creep susceptible polymers are, in ascending order, polyester, polypropylene and polyethylene.

The higher the load, the greater the creep strain and the higher the strain rates, increasing the chances of creep failure.

Muller-Rochholz and Reinhard (1990) noted that the increase in temperature significantly accelerates creep for polypropylene. In polyester, this influence is negligible.

2.3 SOIL BEHAVIOR

Before examining the factors that influence the stress in the reinforcements, a brief review of the behavior of soils will be presented. The actual behavior of soil is non-linear, non-elastic, and the parameters vary according to the stress level. The direct modeling of such behavior would be very complex. In order to address the problem, stress-strain hyperbole model have been developed to be used in non-linear incremental analysis (Duncan et al., 1980). This model was used to develop design methods that will be presented, as follows. Using this procedure the strain calculations corresponding to a particular stress level are not performed in a single step. Instead, the application of the load is divided into several increments.

In each increment of the analysis, it is assumed that the soil has a linear elastic behavior, based on constant parameters (Young's modulus and Poisson's coefficient). Linear behavior means that the soil obeys Hooke's law, in other words, that strains are directly proportional to the applied stresses. Elastic behavior means that all suffered strains are recovered when the load is removed.

In the next increment, however, the values of the elastic parameters are modified. With this simple procedure, it is possible to simulate the actual behavior of the soil, under a certain load, using a linear-elastic model. The condition for the analysis to be accurate is that, for simulation purposes, the load is divided into a sufficiently large number of increments.

Duncan et al. (1980) use the hyperbolic relationship between stress and strain proposed by Kondner and Zelasko (1963), represented in Figure 2.6:

$$(\sigma_1 - \sigma_3) = \cfrac{\varepsilon}{\cfrac{1}{E_i} + \cfrac{\varepsilon}{(\sigma_1 - \sigma_3)_{ult}}}$$

where $(\sigma_1 - \sigma_3)$ is the deviatory stress; $(\sigma_1 - \sigma_3)_{ult}$ is the ultimate deviatory stress of the soil; and E_i is the initial tangent modulus.

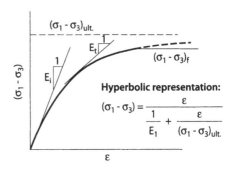

Figure 2.6 *Representation of a hyperbolic stress-strain curve (Duncan et al., 1980)*

For all soils, except for saturated clays tested under non consolidated and non drained conditions, an increase in confining stress σ_3 results in a stress-strain curve with the greater initial tangent modulus and ultimate deviatory stress. The tensile stress "E_i" according to σ_3 is represented by the following equation suggested by Janbu (1963):

$$E_i = K \cdot P_a \cdot \left(\frac{\sigma_3}{P_a}\right)^n$$

Where K is the modulus number in the hyperbolic stress-strain model (dimensionless) and n is the "exponent modulus" in the hyperbolic stress-strain model (dimensionless).

Figure 2.7 shows the E_i variation according to σ_3.

The variation of $(\sigma_1 - \sigma_3)_{ult}$ with σ_3 is given by the relationship $(\sigma_1 - \sigma_3)_{ult}$ to failure deviatory stress, $(\sigma_1 - \sigma_3)_f$, which is related to Mohr-Coulomb failure envelope as follows:

$$(\sigma_1 - \sigma_3)_f = R_f \cdot (\sigma_1 - \sigma_3)_{ult}$$

$$(\sigma_1 - \sigma_3)_f = \frac{2 \cdot c \cdot \cos\phi + 2 \cdot \sigma_3 \cdot \sin\phi}{1 - \sin\phi}$$

where R_f is the failure ratio of the hyperbolic model of Duncan et al. (1980).

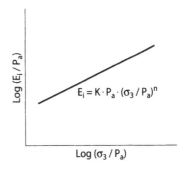

Figure 2.7 *Variation of tangent modulus due to confining pressure (Duncan et al., 1980)*

The inelastic behavior of the soil is represented by the use of different modulus for loading and unloading conditions. In the hyperbolic model of Duncan et al. (1980), the same modulus E_{ur} is employed for unloading and reloading situations, as shown in Figure 2.8. The value of E_{ur} refers to the confining pressure of σ_3 as follows:

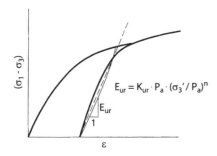

$$E_{ur} = K_{ur} \cdot P_a \left(\frac{\sigma_3}{P_a} \right)^n$$

Where E_{ur} is the tangent loading and unloading modulus (kPa) and K_{ur} is the modulus number for loading and unloading in the hyperbolic stress-strain model (dimensionless).

Figure 2.8 *Unloading and reloading hyperbolic Modulus (Duncan et al., 1980)*

Duncan et al. (1980) state that the $K_{ur} = K$ relationship usually ranges from 1.2 to 3.0.

In Table 2.1, the conservative hyperbolic parameters suggested by

TABLE 2.1 CONSERVATIVE HYPERBOLIC PARAMETERS FOR VARIOUS SOILS (Duncan et al. 1980)

Class. Unif.[1]	CL[2] (%)	$\gamma_m^{(3)}$ (kN/m³)	ϕ (°)	c (kPa)	K[4]	n[5]
	105	24	42	0	600	0.40
GW,GP	100	23	39	0	450	0.40
SW, SP	95	22	36	0	300	0.40
	90	21	33	0	200	0.40
	100	21	36	0	600	0.25
SM	95	20	34	0	450	0.25
	90	19	32	0	300	0.25
	85	18	30	0	150	0.25
	100	21	33	24	400	0.60
SM-SC	95	20	33	19	200	0.60
	90	19	33	14	150	0.60
	85	18	33	10	100	0.60
	100	21	30	19	150	0.45
CL	95	20	30	14	120	0.45
	90	19	30	10	90	0.45
	85	18	30	5	60	0.45

(1) Soil classification according to the unified system; (2) Compaction level AASHTO; (3) unit dry weight

TABLE 2.2 UNIFIED SOIL CLASSIFICATION SYSTEM

Coarse-grained soil (% smaller than sieve #200 < 50)	G: Gravel (% gravel > % sand)	%sieve #200 < 5	GW: **well graded** (CNU > 4 and 1 < CC < 3)
			GP: **poorly graded** (CNU < 4 or 1 > CC > 3)
		%sieve #200 > 12	GC: **clayed** / GM: **silty**
		5 < %sieve #200 < 12	GW-GC, GP-GM, etc.
	S: Sand (% gravel < % sand)	%sieve #200 < 5	SW: **well graded** (CNU > 6 and 1 < CC < 3)
			SP: **poorly graded** (CNU < 6 or 1 > CC > 3)
		%sieve #200 > 12	SC: **clayed** / SM: **silty**
		5 < %sieve #200 < 12	SW-SC, SP-SM, etc.
Fine-grained soil (% smaller than sieve #200 > 50)	C: Clay	CL: medium compressibility	
		CH: high compressibility	
	M: Silte	ML: medium compressibility	
		MH: high compressibility	
	O: Organic	OL: medium compressibility	
		OH: high compressibility	

Duncan et al. are presented (1980) for different soil types that can be used for RSRS design, in case the mechanical behavior parameters of soils are not available. Soils are categorized according to the Unified Classification System (Table 2.2).

It should be noted that soils analyzed by Duncan et al. (1980) are typical of countries with temperate climates, where design criteria recommend avoiding the use of fine-grained sedimentary soils, due to its poor performance. In tropical countries, however, there are lots of fine-grained lateritic or residual origin soils, which have been used successfully for decades in several works, with excellent performance. Table 2.3 presents conservative parameters proposed by Marques, Ehrlich and Riccio (2006)

TABLE 2.3 CONSERVATIVE HYPERBOLIC PARAMETERS FOR BRAZILIAN SOILS
(Marques, Ehrlich, Riccio, 2006)

Class. Unif.	GC (%)	γ_m (kN/m³)	ϕ(°)	c (kPa)	K	n
SM	100	21	36	20	600	0.40
	95	20	34	15	450	0.40
	90	19	32	10	300	0.40
	85	18	30	5	150	0.40
SM-SC	100	21	33	25	700	0.60
	95	20	33	20	500	0.60
	90	19	33	15	350	0.60
	85	18	33	10	300	0.60
ML	100	19	28	25	250	0.70
	95	18	28	20	200	0.70
	90	17	28	15	150	0.70
	85	16	28	10	100	0.70
MH	100	17	25	30	500	0.70
	95	16	25	25	300	0.70
	90	15	25	15	250	0.70
	85	14	25	10	200	0.70

TABLE 2.3 CONSERVATIVE HYPERBOLIC PARAMETERS FOR BRAZILIAN SOILS (Marques, Ehrlich, Riccio, 2006)   *(continued)*

Class. Unif.	GC (%)	γ_m (kN/m³)	ϕ(°)	c (kPa)	K	n
CL	100	19	28	25	200	0.45
	95	18	28	20	170	0.45
	90	17	28	15	140	0.45
	85	16	28	10	100	0.45
CH	100	17	25	30	500	0.50
	95	16	25	25	300	0.50
	90	15	25	20	250	0.50
	85	14	25	10	200	0.50

for Brazilian compacted soils, obtained from the analysis of test results provided by Maiolino (1985) and Melo (1986).

2.4 INFLUENCE FACTORS ON REINFORCEMENT TENSION

Before addressing the determination of the stresses that occur in geosynthetics in an RSRS, it is convenient to study the factors that influence it.

The interaction between soil and geosynthetic depends on the resistance and deformability properties of the soil, geometry, resistance and stiffness of the geosynthetic and the boundary conditions.

2.4.1 Soil-reinforcement interaction mechanisms

The basic soil-reinforcement interaction mechanisms provided by geogrids are different than those presented by other reinforcements such as geotextiles, for example. In geotextiles and geobars, stress transfers occur only due to friction, because of its format.

Because geogrids are hollow reinforcing elements, the soil penetrates the voids, and the interaction between soil and geogrid occurs through shear at the interfaces and passive resistance against the transverse elements. It is generally accepted that shear develops in both longitudinal and transverse elements.

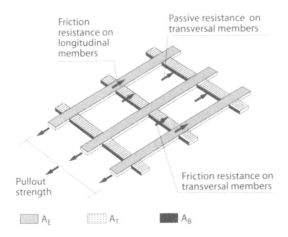

Friction resistance on longitudinal members

Passive resistance on transversal members

Pullout strength

Friction resistance on transversal members

A_E A_T A_B

Figure 2.9 *Soil-geogrid interaction mechanisms (adapted from Wilson-Fahmy and Koerner, 1993)*

Juran and Chen (1988) cite three interaction mechanisms: lateral friction in longitudinal elements, interlocking of the soil trapped between the voids and passive resistance against the transverse elements (Figure 2.9).

It is hard to estimate the portions that belong to each of the mechanisms. The larger the area of longitudinal members, the higher the interface shear resistance portion will be. Bergado et al. (1993) state that, in case of geogrids with thin strips, the portion granted to shear may represent about 10% of the mobilized resistance.

To study the soil-geogrid interaction mechanisms, interface shear or pullout tests are primarily used.

According to Milligan and Palmeira (1987), there are three possible mechanisms for internal collapse in a reinforced soil structure (Figure 2.10). As a first mechanism, the failure surface 1 crosses and eventually leads the reinforcement to fall in tension at point A. If reinforcement failure does not occur at point A, pullout may occur along AB. It is also possible that sliding would occur at the soil-reinforcement interface along CD, according to surface 2.

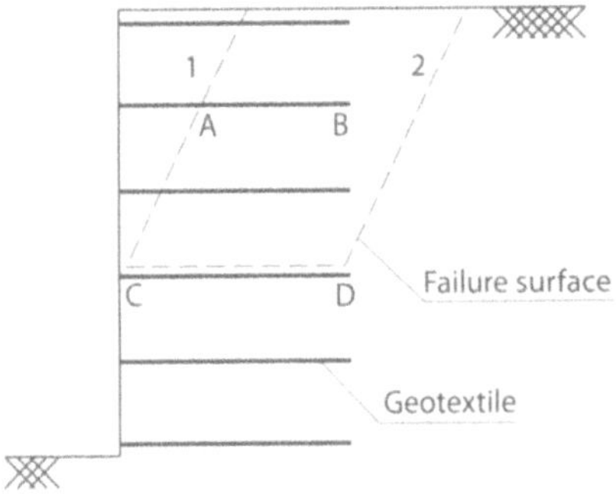

In Figure 2.11, three tests are schematically presented. These tests are capable of inducing, in the reinforcement sample, efforts similar to those that occur in the field: direct shear, direct shear with sloped reinforcement and pullout, which aim to determine the resistance parameters of the soil-geosynthetic interface.

The shear tests consist mostly of adaptations of the traditional direct soil shear test. The shear plane of the boxes should be

Failure surface

Geotextile

Figure 2.10 *Failure mechanisms in a reinforced soil structure (Milligan, Palm, 1987)*

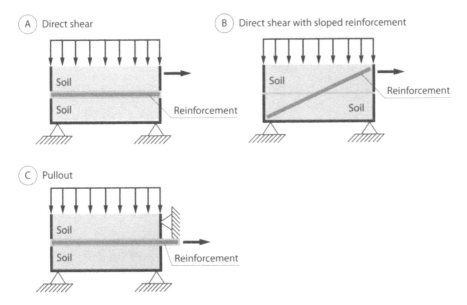

Figure 2.11 *Tests for: (A) direct shear of soil-reinforcement interface, (B) direct shear with inclined reinforcement, (C) pullout of reinforcement*

adjusted to pass exactly at the interface between soil and geosynthetic. It is also possible to place the geosynthetic at an angle to the shear plane surface.

In the pullout tests, the reinforcement is confined by soil in the upper and lower faces. One end is buried in the ground, while the other is attached to a clamp, where increasing tensile strength is applied for pullout.

It should be noted, however, that the boundary conditions vary significantly from test to test and therefore the obtained parameters for interface resistance may vary greatly.

Besides the direct shear and pullout tests, shear tests on an inclined plane or ramp tests can also be performed to determine resistance properties of soil-reinforcement interfaces. The technique has been used by several authors for some decades. Figure 2.12 shows an arrangement recently used by Becker (2001).

Tests should be performed for different vertical stresses, in order to define the shear resistance envelope. The goal is to determine the envelope and its corresponding parameters.

In the active zone, the direction of soil-reinforcement relative movement (and the mobilized stresses) is opposite to that found in the

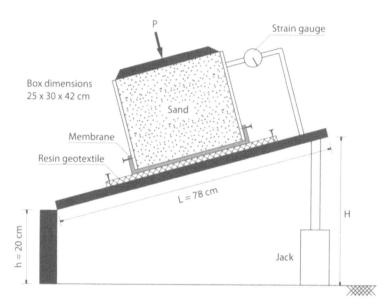

Figure 2.12 *Schematic representation of a ramp test (Becker, 2001)*

resistant area. Thus, the point at which we observe the maximum value of tensile strength in the reinforcement (T_{max}) should occur in the potential failure surface that separates these two zones (Figure 2.13).

The maximum tensile strength value of the reinforcement depends on several factors, particularly the reinforcement and soil stiffness properties and the induced stresses in the soil due to compaction. Based on the work of Ehrlich and Mitchell (1994), a discussion of these factors and how they can be evaluated in a reinforced soil wall design is presented as follows.

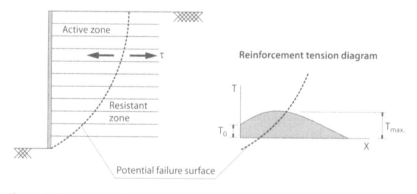

Figure 2.13 *Active and resistant zones, and mobilized tensile load along the reinforcement (Ehrlich, Azambuja, 2003)*

2.4.2 Relative soil-reinforcement stiffness

Under working conditions, it is reasonable to consider the hypothesis of perfect interface adherence between the soil and the reinforcements, i.e., there is no slip between the soil and the reinforcements (Jewel, 1980, Dyer and Milligan, 1984). Thus, the soil and reinforcement strain are the same at this interface.

Under such conditions, the behavior of reinforced soil is independent from the strength transfer mechanism of the soil-reinforcement interface and thus also independent of the type of reinforcement used (geogrids, geotextiles, etc.).

The equilibrium stress or strain between geosynthetics and soil depends on the relationship between the stiffness of both, which is represented by the relative stiffness index (S_i), as defined by Ehrlich and Mitchell (1994):

$$S_i = \frac{J_r}{K \cdot P_a \cdot S_v}$$

where K is the hyperbolic modulus number of the soil (Duncan et al., 1980), P_a is the atmospheric pressure, and S_v is the vertical spacing of the reinforcements.

The reinforcement tensile stiffness modulus, J_r, is determined through the tensile strength per width unit of the geosynthetic (T_r), determined in wide-width strip tests (NBR 12824/93), and the maximum elongation at burst (ε_r), as follows:

$$J_r = \frac{T_r}{\varepsilon_r}$$

Figure 2.14 illustrates a simple model for the mechanism of stress mobilization in a mass of reinforced soil. Two curves are shown representing a soil (a) where the stress-strain curve shows no peak resistance and another (b) with a peak resistance. There are also curves of two reinforcements with different stiffness, $(S_i)_1$ and $(S_i)_2$. Hypothetically, under zero horizontal strain,

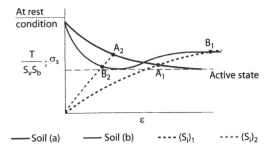

Figure 2.14 *Stress mobilizations in a mass of reinforced soil*

the soil would be in the state of stress corresponding to at rest conditions and there would be no tension load in the reinforcements. With increasing horizontal strain, the horizontal soil stresses ($\sigma_{s,x}$) decrease, approaching the active condition. Simultaneously, the stresses in the reinforcements increase until the equilibrium of the reinforced soil mass is satisfied. This equilibrium can be achieved with relatively small strains, when the reinforcement is stiffer (points A_2 and B_2). When the reinforcement has lower stiffness, the strains required to reach equilibrium are higher (point A_1), soil failure may occur when the soil stress-strain curve shows a post-peak strain softening (point B_1).

Figure 2.15 shows a typical test result for a polyester geogrid, from which it is possible to determine a value for tangent stiffness modulus at failure equal to 1270 kN/m, undermining the initial curvature of the graph.

Table 2.4 shows S_i values for typical reinforcement conditions, considering soil with K $=$ 450 kPa and vertical spacing of 0.5 meter. Soils with different stiffness or other spacing may affect the value of S_i.

It is worth mentioning that the larger the value of S_i, the stiffer the reinforcement and, in general, higher tensions are mobilized by it and lower strains are suffered by the reinforced soil structure. Thus, structures reinforced with PVA geogrids, for example, tend to have higher tensile strengths and lower strain compared to those observed in structures reinforced with nonwoven geotextiles. However, it should be noted that in the case of soils with post-peak strain softening, there may be a different situation. Higher stress mobilization may be required in reinforcements with lower stiffness to reach equilibrium (soil type b, point B1, Figure 2.14).

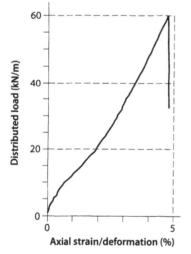

Figure 2.15 *Resultado de um ensaio de tração faixa larga, para uma geogrelha do tipo Fortrac 55/25-20/30 MP*

2.4.3 **Soil compaction**

The effect of soil compaction on the behavior of reinforced masses must be considered, as compaction can significantly affect the internal stresses of these structures.

TABLE 2.4 TYPICAL VALUES FOR THE RELATIVE STIFFNESS
INDEX

Type of geosynthetic	S_i
PVA geogrids [1]	0.020 a 0.200
PET geogrids [1]	0.010 a 0.100
PP geogrids [1]	0.015 a 0.150
Woven geotextiles [1]	0.010 a 0.100
Nonwoven geotextiles in general [2]	0.001 a 0.003

(1): tensile strength between 20 and 200 kN/m
(2): tensile strength between 20 and 50 kN/m

Duncan and Seed (1986) indicate that the compaction operation may be modelled by load and unload cycles that would induce high horizontal residual stresses in the soil. Depending on compaction, the horizontal residual stresses can be much greater than those from geostatic origin, which significantly increases the tensile mobilized load in the reinforcements. That does not mean that compaction is harmful, because of that the structure becomes less sensitive to surcharge loading applications, which induce lesser stresses than those resulting from compaction. The final effect of the process can be understood as a kind of overconsolidation of the soil.

In the field, the soil goes through a complex stress path as a result of the various load and unload cycles caused by the passing of compaction equipment. In the model proposed by Ehrlich and Mitchell (1994), however, the path of stresses is simplified, assuming only one load and unload cycle for each layer of soil, as illustrated in Figure 2.16. The ordinates axis plots the effective vertical stresses and the abscissa axis plots the effective horizontal stresses. Point (1) represents the stress state resulting from the placing of a layer of soil, point (2) represents the stress state during compaction operation. The passing of the compaction equipment leads to increased vertical stress, raising them to $\sigma'_{zc,i}$ (maximum effective vertical stress induced during compaction including loads of dynamic origin). At the same time, there is an increase in horizontal stresses, which reach a maximum value. With the withdrawal of the equipment at the end of the compaction operation, the vertical stress returns to its initial value, σ'_z, as represented by Point (3). However, the same does not occur with the horizontal stress, which suffers a small reduction to a value higher than

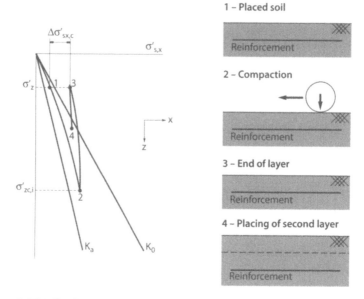

1 – Placed soil

Reinforcement

2 – Compaction

Reinforcement

3 – End of layer

Reinforcement

4 – Placing of second layer

Reinforcement

Figure 2.16 *Effective stress trajectories at a point within the soil mass during the construction of a layered compacted fill*

the original, because the soil is not an elastic material. Thus, the ground keeps a "memory" of the horizontal stress due to compaction ($\sigma'_{sx,c}$). The placing of the next layer will lead to an increase in vertical stress and a small variation in horizontal stress, as represented in (4).

The compaction "memory" will be completely erased only when the vertical stress caused by the layers' own weight above the considered point surpasses the maximum vertical stress induced during the compaction operations, $\sigma'_{zc,i}$.

If there were no compaction, points (1), (2) and (3) of the curve would be coincident, that is, there would be no unloading or residual horizontal stress. In this condition, the horizontal stress value would be lower and it would be between the values corresponding to the active and rest conditions.

In the case of a vibratory tamper, the value of $\sigma'_{zc,i}$ can be estimated as the average vertical stress acting on the soil-plate contact, i.e.,

$$\sigma'_{zc,i} = \frac{Q}{B \cdot L}$$

where Q is the equivalent static load of the compactor (mass times the dynamic amplification factor), B is the width of the plate, and L is the length of the plate.

In the case of compactor rollers, the model presented by Ehrlich and Mitchell (1994) can be used. The authors developed a method to determine stress induced by compaction using the procedure proposed by Duncan and Seed (1986). Compaction is considered a superficial mobile and transient load, with finite lateral extension, and modeled as an equivalent one-dimensional load.

Much as in common field compaction, the layers are relatively thin (up to 30 cm thick); all the soil in a layer can be considered equally compacted.

It is known, however, that the lateral strain of the reinforced soil layer in the direction perpendicular to the face of the wall, reduces the maximum horizontal stress induced by compaction when compared to the maximum stress that would exist in cases where there are no lateral strains. Therefore, the actual maximum horizontal stress induced by compaction is also a function of the stiffness of the reinforcements (point 3 in Figure 2.16).

However, the vertical stress induced by compaction, $\sigma'_{zc,i}$, may be assumed independent from horizontal strains and reinforcement stiffness, and determined for the no lateral deformation condition by:

$$\sigma'_{zc,i} = \sigma'_{xp,i}/K_0$$

where $\sigma'_{zc,i}$ is the maximum vertical stress due to compaction; $\sigma'_{xp,i}$, is the maximum horizontal stress that would be induced by compaction of the soil layer in absence of lateral deformation in derection of reinforcement; and K_0 the pressure coefficient at rest ($K_0 = 1 - \sin \phi'$).

The one-dimensional compaction model is a simplification of the actual behavior. In the near-surface zone under the roller, the soil may be in a state of plastic failure. Considering the movement of the roller parallel to the face and plane strain failure of the soil (Figure 2.17), Ehrlich and Mitchell (1994) developed the following expression for calculating $\sigma'_{zc,i}$, based on the theory of bearing capacity:

$$\sigma'_{zc,i} = (1 - \nu_0) \cdot (1 + K_a) \cdot \sqrt{\frac{1}{2}\gamma' \cdot \frac{Q \cdot N_\gamma}{L}}$$

where K_a is the active pressure coefficient by the formulation of Rankine $= \tan^2(45 - \phi'/2)$; γ is the specific weight of the compacted soil; L is the length of the roller drum, and Q is the maximum vertical operating force of the roller.

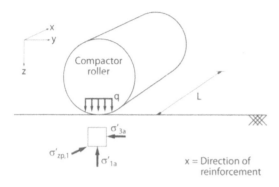

Poisson's coefficient at rest (v_0) can be estimated as if the soil presents elastic linear behavior. Thus, through continuum mechanics, we would have:

$$v_0 = \frac{K_0}{1 + K_0}$$

Figure 2.17 *Stress state of the soil near compactor roller, according to Ehrlich and Mitchell (1994)*

The estimation of the pressure coefficient at rest can be achieved through the correlation of Jaky:

$$K_0 = 1 - \sin \phi'$$

The capacity coefficient (N_γ) can be determined by:

$$N_\gamma = \tan\left(45° + \frac{\phi'}{2}\right) \cdot \left[\tan^4\left(45° + \frac{\phi'}{2}\right) - 1\right]$$

Table 2.5 shows the characteristics of various vibrating rollers and Table 2.6, the characteristics of compaction vibrating tampers, as stated in the manufacturers' catalogs. The centrifugal forces listed are the maximum vibration amplitude of the rollers. Figure 2.18 shows the $\sigma'_{zc,i}$ values of compactor rollers for soil with specific 18 kN/m³ weight and various angles of friction. For soil with a different unit weight (γ_2), simply multiply the value obtained from $\sigma'_{zc,i}$ by $\sqrt{\gamma_2/18}$.

The compaction may be considered a kind of overconsolidation of the soil. While the vertical stress induced by compaction ($\sigma'_{zc,i}$) is greater than the vertical stress due the soil's self weight above the considered layer (σ'_z), the horizontal stress would be the one induced by compaction. When σ'_z exceeds $\sigma'_{zc,i}$, the effects of compaction are no longer felt by the soil and the horizontal stress is controled by the overburden stress. Thus, the maximum past vertical stress including compaction at the end of

TABLE 2.5 CHARACTERISTICS OF VARIOUS COMPACTOR VIBRATING ROLLERS
(adapted from Case, 2007; Autramaq, 2007a and 2007b and Dynapac, 2007)

Manufacturer	model	Roller weight (kN)	Roller width (m)	Equivalent static load (kN)	Vertical stress/stress (kPa)
CASE®	SV216	99.5	2.20	325	*
	SV212	72.3	2.20	277	*
	SV208	39.4	1.70	145	*
	DV201	12.3	1.00	29	*
MÜLLER®	VAP55P	–	1.68	190	*
	VAP70P	–	2.14	320	*
DYNAPAC®	CA134PD	19.6	1.37	89	*
	CA150PD	39.2	1.68	143	*
	CA250PD	72.6	2.13	300	*
	CA500PD	101.0	2.13	300	*

* See Figure 2.19.

TABLE 2.6 CHARACTERISTICS OF COMPACTION FROG-TYPE VIBRATING TAMPERS
(adapted from Dynapac, 2007 and Wacker, 2007)

Manufacturer	model	Equivalent static load (kN)	Base area (m²)	Vertical stress/stress (kPa)
DYNAPAC®	LT500	10.0	0.076	132
	LT600	14.8	0.092	160
	LT700	18.6	0.092	201
WACKER®	BS 50-4	14.7	0.092	159
	BS 60-4	15.6	0.092	169
	BS 70-2i	17.8	0.092	193

construction (overconsolidation pressure), σ'_{zc}, is defined as

$$\sigma'_z \leqslant \sigma'_{zc,i} \quad \Rightarrow \quad \sigma'_{zc} = \sigma'_{zc,i}$$
$$\sigma'_z > \sigma'_{zc,i} \quad \Rightarrow \quad \sigma'_{zc} = \sigma'_z$$

In Figure 2.19, the influence of soil compaction on depth is represented. The depth influence of compaction, z_c, is given by:

$$z_c = \frac{\sigma'_{zc,i}}{\gamma'}$$

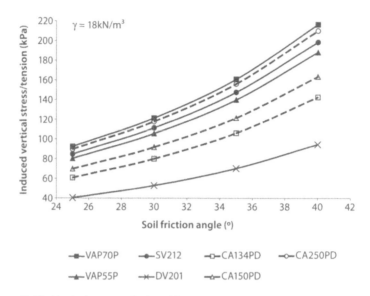

Figure 2.18 *Vertical stresses induced by several compactor rollers*

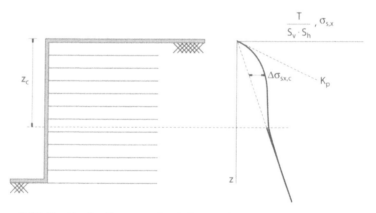

Figure 2.19 *Depth of soil compaction influence (Ehrlich, Azambuja, 2003)*

Compaction is the major factor determining the internal stresses for depths lower than z_c. For typical conditions, z_c can reach about 6 meters for reinforced soil walls with vertical or sub-vertical faces (Ehrlich, Mitchell, 1994), and up to 10 meters for reinforced embankments with slopes lesser than 70° (Dantas, 1998). $\Delta\sigma'_{sx,c}$ represents the increase in horizontal stress - beyond that of merely geostatic origin – caused by soil compaction.

Analysis and design 3

3.1 Introduction

A reinforced soil structure must meet internal and external stability conditions. Its design specifications consist of several steps for defining dimensions, materials and verification of safety factors, as will be described below.

3.2 External stability

The analysis of external stability can be conducted considering the reinforced soil mass as a conventional gravity wall. This "wall" guarantees the stability of the no reinforced soil zone. Under the earth pressure of the non-reinforced mass (E), walls are designed in order to ensure stability to avoid sliding and overturning, beyond the load capacity of the foundations, and prevent global instability (Figure 3.1).

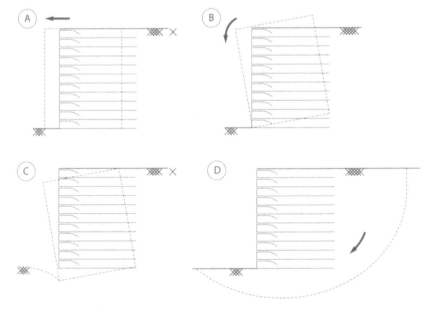

Figure 3.1 *Mechanisms of external instability: (A) sliding, (B) overturning, (C) load capacity of foundation; (D) global instability*

To determine the ground pressure that the non-reinforced soil mass imposes on the reinforced mass (E), it is possible to adopt the classical theories based on limit equilibrium. Many authors recommend using the Coulomb formulation, assuming the friction between the wall (reinforced area) and the ground (non-reinforced area) are equivalent to the angle of internal friction of the soil at critical state ($\delta = \phi_{cv}$). However, some authors consider the mobilization of friction unlikely, since the reinforced area does not behave as a stiff block. Thus, the formulation of Rankine would be more appropriate and the active pressures would be parallel to the ground surface ($\delta = 0$).

If there is intention to apply external overloads on the embankment of the retaining structure, the resulting stress increases (Section 3.9) must be considered in the external stability analysis.

Just as the stability analysis of conventional walls, the verification of external stability can be achieved through the adoption of global safety factors (deterministic approach), taking into account the slope of the structure face (ω) and the material of the foundation or by using weighting factors (probabilistic approach). In the event that deterministic criteria are adopted, the external stability takes into account the safety factors and mechanical conditions listed in Table 3.1 (Elias, Christopher; Berg, 2001).

TABLE 3.1 SAFETY FACTORS AND MECHANICAL CONDITIONS

Verification	Safety factor	Note
Sliding	$\geqslant 1.5$ $\geqslant 1.3$	Structures with $\omega > 65°$ Structures with $\omega < 65°$
Overturning	$\geqslant 2.0$	Structures with $\omega > 65°$
Bearing capacity of foundation	$\geqslant 2.5$	
Global stability	$\geqslant 1.5$ $\geqslant 1.3$	Critical conditions Non critical conditions
Seismic stability	FS equivalent to 75% of FS values can be used for static analysis	
Verification	**Mechanic conditions**	**Note**
Eccentricity	$e \leqslant B/6$ $e \leqslant B/4$	soil supported structures rock supported structures

3.3 INTERNAL STABILITY

The particular aspect in design of a reinforced soil wall is the internal stability analysis. The internal design of a RSRS must ensure safety against rupture and pullout of the reinforcements, and avoid instability located on the face. For this purpose, the following verifications are usually performed:

3.3.1 Maximum tension in the reinforcements

The determination of the maximum tension in the reinforcements (T_{max}) is a key aspect in the internal stability analysis. To avoid reinforcements failure, the T_{max} value should not be greater than the minimum expected value for the geosynthetic design resistance, considering the environmental conditions and installation, and a reference time for the project (t_d), ensuring an adequate safety factor (Figure 3.2A).

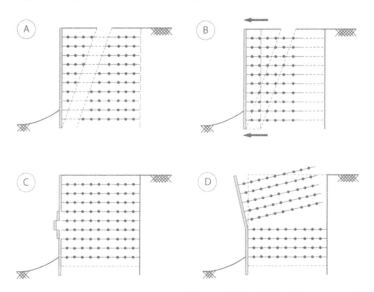

Figure 3.2 *Mechanisms for external stability analysis: (A) reinforcement failure, (B) reinforcement pullout, (C) detachment of the face, (D) local instability (Ehrlich, Azambuja, 2003)*

3.3.2 Pullout resistance

The design of a reinforced soil wall must also ensure minimum embedding of the reinforcement in the resistance zone, thus preventing its pullout (Figure 3.2B). Therefore, the T_{max} value should not exceed the pullout resistance value of the reinforcement embedment into the resistant zone

(P_r), in the respective layer, ensuring the safety factor. It is also necessary to know the position of the critical surface that defines the boundary between the resistant and active zones.

3.3.3 Reinforcement and facing connections

Despite being crucial to the final aspect of the structure, the facing system does not play a relevant role in internal stability of the reinforced soil mass. Its function, from a mechanical viewpoint, is to ensure the stability of the soil masses located between the geosynthetic layers in areas close to the face. Global equilibrium would be possible even in the absence of facing.

However, the connection between the reinforcement and the face should ensure that the remaining existing stresses in the reinforcement are transferred to the face. In general, the connections have less resistance than the reinforcements, but, to compensate, the mobilized tensile force in the reinforcement close to the face (T_0) is lesser than T_{max} (Figure2.14). In an appropriate reinforced soil retaining system, the permissible connection resistance ($P_{r,0}$) must be greater than T_0, avoiding the detachment of reinforcements at the face (Figure 3.2C).

3.3.4 Local failure

Finally, as with any incremental retaining system, the instability in any given wall level should also be verified, as shown in Figure 3.2D.

3.3.5 Safety factors for internal stability

In the event that deterministic criteria are adopted, the internal stability should meet the safety factors listed in Table 3.2 (Elias, Christopher; Berg, 2001)

TABLE 3.2 SAFETY FACTORS AND MECHANICAL CONDITIONS

Condition	Verification	Safety factor	Note
Reinforcement failure	$T_d \geqslant T_{max} \cdot FS$	$\geqslant 1.50$	permanent and critical works
		$\geqslant 1.15$	temporary and non critical works
Pullout	$P_r \geqslant T_{max} \cdot FS$	$\geqslant 1.50$	
Connection stability	$P_{r,0} \geqslant T_0 \cdot FS$	$\geqslant 1.50$	

3.4 DETERMINATION OF MAXIMUM REINFORCEMENT TENSION

The determination of the maximum tension (T_{max}) that acts on each level of the reinforcement is one of the most important aspects to be considered in a reinforced soil structure project, and the usual procedures for obtaining it are based on limit equilibrium methods. The T_{max} value is determined by considering the necessary strengths to reach local equilibrium; in this case, the tensile strength of the reinforcement and soil shear resistance (Leshchinsky; Boedeker, 1989; Jewell, 1991; Elias, Christopher; Berg, 2001).

This type of procedure is limited, as it does not consider the reinforcement stiffness and compaction effects in the analysis. Despite the advantages of its simplicity, according to Abramento and Whittle (1993), limit equilibrium methods are not reliable for estimating the magnitude and distribution of stresses in the reinforcements under working conditions.

Despite cases studies where it has been demonstrated that these methods may lead to a conservative design, Alexiew and Silva (2003) and Silva and Vidal (1999) call atention for possible conditions against safety, under certain conditions of face sloping or in cases of failure surfaces that move beyond the reinforced soil zone.

To overcome the above stated shortcomings of the limit equilibrium methods, some procedures have been developed under working conditions and they consider, more realistically, the complex behavior of reinforced soil structures.

3.4.1 Analysis methods using limit equilibrium

These are the most widely disseminated and used methods, probably due to the ease of use and designers' familiarity with the concepts. These methods adopt the following hypotheses:

- Structure in a state of imminent collapse;
- Shape and location of failure surface;
- Rigid, perfectly plastic behavior for the soil;
- Inclination and distribuition with depth of the mobilized tension in the reinforcements;
- Full mobilization of shear resistance of the soil along the failure surface.

The hypotheses from limit equilibrium methods are major limitations, considering that the reinforced soil structures generally work far from failure conditions, the potential failure surfaces are not well known, the behavior of soil is non-linear elasto-plastic and shear resistance is mobilized in an uneven manner over the potential surface due to the compatibility of soil-reinforcement strains.

Some limit equilibrium methods satisfy equilibrium of stresses, usually achieved by adopting a distribution of earth pressures that must be equilibrated by the geosynthetics' tensile strengths.

Other methods adopt circular or spiral failure surfaces, as in slope stability analysis, and consider the tensile strengths of the reinforcements in calculations of stability by equilibrium of strengths and moments.

In both cases, the disregard for soil and reinforcement strains, as well as the effects of compaction and soil-reinforcement relative stiffness often leads to gross errors in design, including disregard for safety. Peralta (2007) reports the case of a monitored prototype reinforced soil wall that presented very small strains on reinforcements, indicating that the acting tensile strengths were negligible. The retro-analysis concluded that the slope was marginally stable even without the reinforcements. The limit equilibrium methods employed in the project, however, indicate high values for tensile strength in the reinforcements.

This work does not have the intention to present a complete review of the available methods or to exhaust the subject, but merely illustrate the main concepts. To do so, the Steward, Williamson and Mohney (1997) method will be briefly explained below. This method is also known as the US Forest Service method. It is a rather simple and much utilized method, also recommended by North American government agencies, such as the US Army Corps of Engineers and the US Forest Service.

The face may be vertical or sloped, and it is assumed that a potential failure surface goes through the reinforced soil mass, with an inclination of $45 + \phi'/2$ from the horizontal as shown in Figure 3.3, dividing the mass in an "active" zone and another "resistant" zone.

To determine the spacing of the reinforcement layers, it is assumed that, if there is no surcharge loading, the horizontal stress distribution in the soil with depth is linear, as follows:

$$\sigma'_x = K \cdot \gamma \cdot z$$

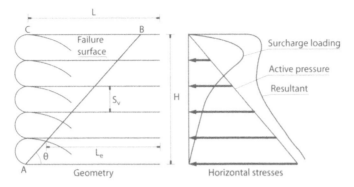

Figure 3.3 *Configuration and mobilized stress in a reinforced soil wall, according to Steward, Williamson and Mohney (1977)*

where σ'_x is the effective horizontal stress in the soil; K is the earth pressure coefficient; γ is the unit weight of the soil, and z is the considered depth.

The method of Steward, Williamson and Mohney (1977) assumes earth pressure corresponding to at rest conditions. Other authors, however, adopt the same method, considering the soil at active condition, and therefore $K = K_a$, furthermore proposing corrections to the value of K to take into account the inclination of the face, for example.

The next step of the method is to determine the spacing between the reinforcements or its equivalent, the mobilized tension in the reinforcements, because they are independent factors. The mobilized tension in any reinforcement is the product of the horizontal stress by the spacing between reinforcements. Thus, for each layer of reinforcement, different spacing may be defined, but the most common is to establish spacing according to the lower layer, where the horizontal stress is greater, and to the resistance of the reinforcement:

$$S_v = \frac{T_d}{\sigma'_x \cdot FS_g}$$

where S_v is the vertical spacing between the reinforcements; T_d is the design geosynthetic tensile load, and FS_g is the safety factor of the structure, usually between 1.3 and 1.5. The design tensile load is determined from the tensile strength of the geosynthetic, considering reduction factors that take into account the effects of installation damage, creep and biochemical degradation. Then, the embedment length required for each reinforcement is determined by:

$$L_e = S_v \cdot \sigma'_x \cdot \frac{FS_a}{2 \cdot (a + \gamma \cdot z \cdot \tan \delta)}$$

where L_e is the length of embedment beyond the potential failure surface; FS_a is the pullout safety factor usually assumed between 1.3 and 1.5; a is the adhesion of the soil-geosynthetic interface; δ is the friction angle of the soil-geosynthetic interface.

The total length of the geosynthetic should also consider the "active" zone and the envelopement, if any, and it is generally addopted constant along the height of the wall. The method presented is basic and simplified. Other more sophisticated considerations can be used for pullout resistance - for example, taking the mobilized loads in the transverse geogrid elements into account - and for calculating the necessary envelopment length or the resistance required in connections with the facing. However, whatever the sophistication incorporated, whether in estimates of horizontal stresses or pullout resistance, the method philosophy is basically the same because there is no consideration of strain compatibility.

3.4.2 Empirical design methods

Elias, Christopher and Berg (2001) present an empirical design procedure endorsed by the Federal Highway Administration (FHWA). The soil and reinforcement stiffness and the effect of compaction are indirectly considered, through the use of the graph shown in Figure 3.4.

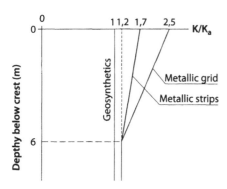

Figure 3.4 *Variation of the K/Ka ratio with depth, for reinforced soil walls, according to Elias, Christopher and Berg (2001)*

The method is very simple to use, but has flaws. In terms of stiffness, difference are made only between metallic or geosynthetic reinforcements. The differences in stiffness between the various types of geosynthetics that, as already noted, can be great, are disregarded. A similar problem occurs in terms of compaction, considered only for metallic reinforcements. The method does not take into consideration the very different energy of the available compaction equipment, which can also vary greatly, as seen previously.

The tensile load in reinforcements at a certain depth of the wall is determined based on the vertical stress at the same depth and the coefficient of lateral pressure, according to the expression:

$$T(z) = \sigma'_x(z) \cdot S_v = \sigma'_z(z) \cdot K \cdot S_v$$

where $T(z)$ is the tensile load at depth z, measured from the crest, and $\sigma'_z(z)$ is the vertical stress at depth z.

To determine the lateral pressure coefficient, the active earth pressure coefficient is multiplied by the K/K_a ratio obtained from the graph, for the desired depth. If geosynthetic reinforcements are being used, the K coefficient would be equal to K_a.

3.4.3 Strain compatibility degisn methods

To overcome the deficiencies of the limit equilibrium methods, some authors have proposed methods based on working stress conditions, such as Adib (1988), Abramento and Whittle (1993), Ehrlich and Mitchell (1994), Dantas and Ehrlich (2000b), among others.

These procedures consider the stress-strain behavior of the reinforced soil mass, including, necessarily, constitutive equations for modelling the soil, reinforcements and, in some procedures, the soil-reinforcement interface. In this work, the method of Ehrlich and Mitchell (1994) and its generalization proposed by Dantas and Ehrlich (2000b) will be presented.

The method is based on the strain compatibility in the reinforcement and the soil, considering the influence of soil-reinforcement relative stiffness and compaction energy. The constitutive model for the reinforcement is linear elastic and assumes that there is no slip between soil and reinforcement.

Each layer of reinforcement is responsible for local horizontal equilibrium of the corresponding horizontal slices of the active zone of thickness (S_v). The first equilibrium condition is therefore:

$$T_{max} = S_v \cdot (\sigma'_x)_{ave}$$

where $(\sigma'_x)_{ave}$ is the horizontal average tension in the influence range of the reinforcement in question.

Figure. 3.5 schematically shows the adopted internal equilibrium hypothesis. The model also assumes that the shear stress at the interface of adjacent slices of soil are null.

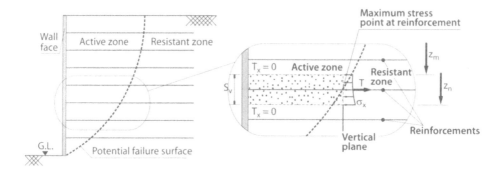

Figure 3.5 *Internal equilibrium of reinforced soil mass, according to Ehrlich and Mitchell (1994)*

The constitutive model adopted for the soil is a modification of the hyperbolic model proposed by Duncan et al. (1980).

Compaction modeling is carried out as described in section 2.4.3, and the maximum horizontal stress induced by compaction is:

$$\sigma'_{xp,i} = \nu_0 \cdot (1 + K_a) \cdot \sqrt{\frac{1}{2}\gamma' \cdot \frac{Q \cdot N_\gamma}{L}}$$

The maximum tensile strength acting on any reinforcement layer, for the final construction condition corresponds to:

$$T_{max} = S_v \cdot \sigma'_{xr} = S_v \cdot K_r \cdot \sigma'_z$$

where T_{max} is the maximum tension in the reinforcement at the end of construction; K_r is the residual earth pressure coefficient at the end of construction; and σ'_z is the vertical stress at the point of interest, at the end of construction. The K_r value can be obtained by iterative solution of the following equation:

$$\frac{1}{S_i} \left(\frac{\sigma'_z}{P_a}\right)^n = \frac{(1 - \nu_{un}^2)\left[(K_r - K_{\Delta 2}) - (K_c - K_{\Delta 2}) \cdot OCR\right]}{\left(\frac{K_u}{K}\right)(K_c \cdot OCR - K_r) K_r^n}$$

where ν_{un} is Poisson's ratio for unloading, from the condition at rest, given by $K_{\Delta 2}/(1 + K_0)$; $K_{\Delta 2}$ is the at rest decremental lateral pressure coefficient for unloading; K_c is the lateral earth pressure coefficient during compaction; OCR is the overconsolidation ratio, given by σ'_{zc}/σ'_z; K_u is the modulus number of the hyperbolic model of Duncan et al. (1980) for unloading; K is the modulus number of hyperbolic model of Duncan et al. (1980); n is the modulus exponent of the hyperbolic model of Duncan et al. (1980).

The values of $K_{\Delta 2}$ and K_c are given by:

$$K_{\Delta 2} = K_0(OCR - OCR^{0.7 \sin \phi'})/(OCR - 1)$$

$$\frac{1}{S_i}\left(\frac{\sigma'_{zc}}{P_a}\right)^n = \frac{(1 - v_0^2)(1 - K_{aa})^2 (K_0 - K_c) \cdot K_0}{(K_c - K_{aa})(K_0 - K_{aa}) \cdot K_c^n}$$

$$K_{aa} = K_a/\{(1 - K_a)[(c'/(\sigma'_{zc} \cdot K_c \cdot \tan \phi') + 1)/R_f] + K_a\}$$

This design procedure can be used in analytical form by or by means of simple nondimensional charts, Ehrlich and Mitchell (1994) and Dantas and Ehrlich (2000a) developed the charts shown in Figure 3.6 and in Figures A.1, A.2, A.3 and A.4 of the Appendix. Calculations are performed considering, for each layer, the vertical stress (σ'_z), the maximum past vertical stress, including compaction (σ'_{zc}) and the β value, which is a parameter that reflects the deformability of the reinforcements.

$$\beta = \frac{\left(\frac{\sigma'_{zc}}{P_a}\right)^n}{S_i}$$

As S_i is a function of the cross-sectional reinforcement area, an iterative process must be adopted in the determination of T_{max}. Typically, three iterations are necessary. In the case of geosynthetics under typical fill conditions and spacing of reinforcements, usually $8 < \beta < 300$. More extensible reinforcements have higher values for β.

For the condition of non-compacted fill, or for depths in which $\sigma'_{zc,i}$ is lesser than the geostatic stress:

$$\sigma'_{zc} = \sigma'_z$$

In all other cases, the compaction effect prevails, and:

$$\sigma'_{zc} = \sigma'_{zc,i}$$

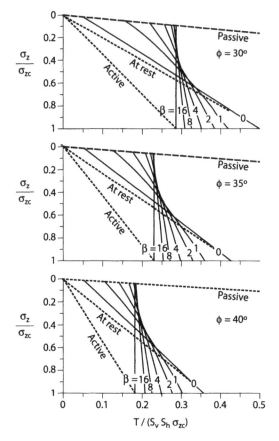

Figure 3.6 *Chart for determination of "χ" for the calculation of T$_{max}$ in structures with vertical faces (Ehrlich, Mitchell, 1994)*

The vertical stress on each layer at the end of construction, (σ'_z) can be determined by the method of Meyerhof (1955), considering the eccentricity of the resulting horizontally based stresses. The calculation is carried out based on the mass equilibrium above the reinforcement layer in question, on its own weight and the active earth pressure exerted by the fill on the wall. For no surcharge loading, σ'_z can be determined by:

$$\sigma'_z = \frac{\gamma' \cdot z}{1 - \left(\frac{K_a}{3}\right) \cdot \left(\frac{z}{L_r}\right)^2}$$

where L_r is the length of the reinforcements and K_a is the active pressure coefficient by Rankine.

In the case of evenly distributed surcharge loading on the embankment (q), equivalent depth (z_{eq}) that considers the presence of the surcharge loading can be given by:

$$z_{eq} = z + \frac{q}{\gamma'}$$

The maximum tensile strength in the reinforcement (T_{max}) can be determined from the dimensionless parameter "χ", obtained from the chart:

$$\chi = \frac{T_{max}}{S_v \cdot S_h \cdot \sigma'_{zc}} \quad \Rightarrow \quad T_{max} = \chi \cdot S_v \cdot S_h \cdot \sigma'_{zc}$$

For a preliminary assessment of the T_{max} value, in the absence of soil deformability tests, the conservative parameters suggested in Tables 2.1 or 2.3 can be used.

Dantas and Ehrlich (1999) present charts that allow for the consideration of cohesion in T_{max} calculations. Figures A.5, A.6 and A.7 of the Appendix provide, respectively, the charts for structures with vertical facing, 3V:1H inclination and 2V:1H inclination.

As expected, cohesion can significantly reduce the active stresses on the reinforcements. Note that the importance of cohesion is greater for less rigid reinforcements. Failure of reinforcements is verified when T_{max}, increased by a safety factor, is lesser than the design geosynthetic tensile load (T_d). T_d value determination is presented in section 3.8.

3.4.4 Comparision between measurements and predictions

Because of the different hypotheses adopted, the methods based on limit equilibrium and strain compatibility offer conflicting results.

Table 3.3 presents, in summary form, a comparison between estimates obtained by the method of Ehrlich and Mitchell (1994) and measured results in three monitored RSRS built in Brazil. On the table, there are some of the RSRS characteristics that were analyzed and the values of the ratios between the estimated and measured maximum tensile strengths.

TABLE 3.3 COMPARISON BETWEEN ESTIMATED AND MEASURED MAXIMUM TENSILE
LOADS (adapted from Peralta, 2007)

RSRS	Characteristics	$T_{estimated}/T_{measured}$ Ehrlich and Mitchell (1994)
Becker (2006)	Height: 4.5 m Face inclination: 1(H) : 5(V) Reinforcement: geogrid Spacing: variable 0.4 m to 0.6 m Length: 4.2 m Compaction equip: Dynapac CA25 roller compactor Soil: residual clayey silt	Min = 1.3 Max = 2.0 Average = 1.6
Benjamin (2006)	Height: 4 m Face inclination: 1(H) : 5(V) Reinforcement: nonwoven geotextile Spacing: 0.4 m Length: 3 m Compaction equip: Wacker, BPS1135W vibrating plate Soil: medium and coarse sand	Min = 0.3 Max = 2.1 Average = 1.5
Riccio (2007)	Height: 4.2 m Face inclination: vertical Reinforcement: geogrid Spacing: variable 0.4 m to 0.6 m Length: 4 m Compaction equip: Dynapac CA25 roller compactor Soil: residual sandy clay	Min = 1.2 Max = 2.6 Average = 2.1

The method of Ehrlich and Mitchell (1994) provided results above those measured in the field, i.e., on the safe side, in 93% of the cases, proving to be reliable. The dispersion of results was relatively small, with the $T_{estimated}/T_{measured}$ ratio ranging between 0.3 and 2.6 (1.7 average).

Peralta (2007) also evaluated ten limit equilibrium methods. The results indicated that, by not considering the compaction effects or the cohesion of the fill, the methods based on limit equilibrium had greater dispersions, overestimating the tensile loads measured in some cases and underestimating them in others.

Dantas (2004) analyzed the behavior of RSRS under working conditions by means of analytic and computational studies. In all numerical simulations, a finite element program, which incorporated the hyperbolic formulation of Duncan et al. (1980) and modeling of compaction according to Seed and Duncan (1986) was used. The results were compatible with the analytical method of Ehrlich and Mitchell (1994). Figure 3.7 presents comparative results of maximum tensile loads on the reinforcements of RSRS with relative stiffness index of $S_i = 0.1$, height of 10 meters, soil friction angle of 30° and variable cohesion values.

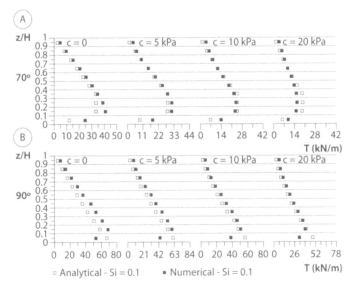

Figure 3.7 *Maximum tensile loads on reinforcements for structures reinforced with geosynthetics, with different values of soil cohesion and inclination: (A) 70° and (B) 90° (Dantas, 2004)*

3.5 DETERMINATION OF THE LOCATION OF REINFORCEMENT MAXIMUM TENSION POINT

The maximum tension is located at the intersection with the potential failure surface that separates the active and resistant zones. In Figure 3.8, conventionally adopted hypotheses for defining this surface are presented (Christopher et al., 1990). For more deformable reinforcements (geotextiles and PET or HDPE geogrids), it is customary to consider the position of T_{max} coincident with the critical surface estimated by Rankine (Figure 3.8A). For less extensible reinforcements (polyaramid geogrids or geobars, for

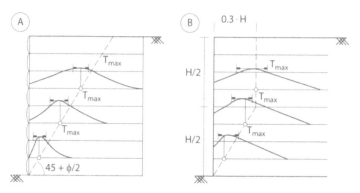

Figure 3.8 T_{max} *location for reinforced soil walls (Christopher et al., 1990): (a) extensible reinforcement, (B) non extensible reinforcement*

example), the system's restrictions to lateral strains cause the critical surface to present itself in a more vertical position. In such cases, the acting position of T_{max} is approximately what is indicated in Figure 3.8B.

Figure 3.9 presents the procedure for establishing the maximum tension location for reinforcements on slopes, suggested by Dantas and Ehrlich (2000a). This procedure was developed based on numerical analysis and can be adopted for reinforcements of any stiffness, with reasonable accuracy . Similar results were obtained through studies in centrifuges by Zornberg, Sittar and Mitchell (1999). A discussion of these results is presented in Dantas and Ehrlich (2000b). In Figure 3.9, we have:

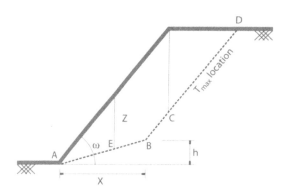

Figure 3.9 T_{max} *location point for reinforcement of steep slopes (Dantas, Ehrlich, 2000a).*

$$\text{for } 45° \leqslant \omega \leqslant 65° \quad \Rightarrow \quad x = \frac{0.75 \cdot H}{\tan \omega} \quad \therefore \quad h = \frac{x}{3}$$

$$\text{for } 65° \leqslant \omega < 90° \quad \Rightarrow \quad x = \frac{0.8 \cdot H}{\tan \omega} \quad \therefore \quad h = \frac{x}{2}$$

3.6 PULLOUT ANALYSIS

The pullout resistance mechanism of geotextiles is based on friction between soil and reinforcement. For geogrids, however, there is an important component of passive resistance in the soil against the transverse elements, in addition to friction.

Due to the deformability of the reinforcement, the application of pullout strength results in a decreasing displacement distribution along the length. The strain is maximum at the beginning of the resistant portion and decreases towards the end of the reinforcement, and may be null. As a consequence of variable strains/deformations, the mobilization of the tangential stress and, therefore, of pullout resistance, is uneven along the reinforcement (Ehrlich, Azambuja, 2003).

Several authors have proposed formulations for estimating geosynthetic reinforcement behavior under pullout efforts (Jewell et al., 1984, Christopher et al., 1990; Bergado; Chai, 1994; Teixeira, 2003). There is, however, a large dispersion of estimated values in relation to measured values in pullout tests, which indicates that the subject is not completely understood. For this reason, the current practice is to adopt conservative estimates.

The conservative procedure suggested by Christopher et al. (1990) for the determination of pullout resistance per cross sectional length unit of the reinforcement (P_r) is:

$$P_r = 2F^* \cdot \alpha \cdot \sigma'_z \cdot L_e \quad \geqslant \quad FS \cdot T_{max}$$

where L_e is the length of the reinforcement in the resistant zone (beyond the potential failure surface); F^* is the pullout resistance factor; α is the scale effect correction factor; and σ'_z is the effective vertical stress in the soil-reinforcement interface.

The pullout resistance factor (F^*) can be more accurately determined by means of pullout tests, considering the soil to be used in the project. Alternatively, a simplified procedure for determining F^* can be used:

$$F^* = f_a \cdot \tan \phi$$

where f_a is the adhesion ratio and ϕ is the soil friction angle.

Table 3.4 presents conservative values of α and f_a.

TABLE 3.4 PARAMETERS FOR PULLOUT ANALYSIS

Type of geosynthetic	adhesion ratio (f_a)	scale factor (α)**
Geogrids	0.8 a 1.0 *	0.7 a 1.0
Geotextiles	0.7 a 0.8	0.6 a 0.8

* According to mesh openings and thickness of transverse members
** In the absence of pullout tests, $\alpha = 0.6$ is adopted

3.7 REINFORCEMENT AND FACING CONNECTIONS

Due to the mobilization of tangential stresses within the active zone, the stress transferred from the reinforcement to the facing is less than T_{max}. This allows the facing to be slim, one of the reasons for the economic competitiveness of reinforced soil systems.

The reduction in stress transfer to the facing depends on the relative soil-reinforcement stiffness, on the stresses induced by compaction and on relative soil-facing stiffness.

In general, stiff reinforcements transfer greater T_{max} portions to the face than those that are deformable. Similarly, facings that are stiffer, tend to attract more stress to the face than flexible facings. Among all these factors, the influence of soil compaction is the most relevant, reason why there is an overall guideline for preventing the passing of heavy compaction equipment close to the facing.

The general expression to analyze the permissible resistance of the reinforcement-facing connection ($P_{r,0}$) is determined according to the type of connection. In wrap-around systems, the anchor of the extremities can be calculated assuming that T_0 is 50% of T_{max}:

$$P_{r,0} = 2F^* \cdot \alpha \cdot \sigma_z' \cdot L_0 \quad \geqslant \quad FS \cdot T_0$$

where L_0 is the envelope anchor length; F^* is the pullout resistance factor; α is the scale effect correction factor; σ_z' is the effective vertical stress at the soil-reinforcement interface along the face; and T_0 is the tensile strength on the reinforcement at the facing.

In rigid facing systems or interlocking block systems, the resistance of the connection between the face and the reinforcement depends on the type and efficiency of the coupling (CR), a parameter that means which percentage of the permissible resistance of the reinforcement-facing connection ($P_{r,0}$) can be supported. Since the efficiency of the coupling depends on the combination of a specific facing system (block or panel) with a specific type of geosynthetic (rigid geogrid, flexible geogrid, woven geotextile etc.), the CR may only be determined by full-scale testing. Due to the difficulty of such tests, the CR values should be informed by the supplier of the block-geosynthetic system. Thus, for connection with blocks or panels, we have:

$$P_{r,0} = CR \cdot T_d \quad \geqslant \quad FS \cdot T_0$$

It should be stressed that, in walls with interlocking block facing, panels or monolithic walls, the value of T_0 is difficult to determine. Therefore, if smaller strains/deformations in the facing are desired, it is recommended that the designer adopt conservative T_0 values (between 80% and 100% of T_{max}).

3.8 DESIGN RESISTANCE OF REINFORCEMENTS
The design strength (T_d) is the one estimated for the reinforcement at the end of a given reference time (service life) for a particular installation environment and a specific loading condition, as stated previously.

Besides the phenomena of environmental degradation and damage that may occur during installation, the polymers are visco-elastic materials, and therefore creep plays an important role in long-term resistance.

In general, the design strength is determined from the following expression:

$$T_d = \frac{T_k(t)}{f_d \cdot f_a}$$

where $T_k(t)$ is the characteristic resistance (long term) at design reference time "t", f_d is the factor for resistance reduction due to mechanical damage by installation ; and f_a is the factor for drag reduction by biochemical degradation.

The characteristic long-term resistance $T_k(t)$ depends on the intensity, duration and nature of load, the type of polymer and temperature of the design. Therefore, its determination should be based on the results of extended duration tension tests (at least 10 thousand hours), using the isochronous diagrams for strength estimation, according to section 2.2.6, which should be provided by the reinforcement manufacturers.

The extended duration tests are performed at a 20°C temperature, which is reasonably consistent with the average temperature inside the walls of reinforced soils. However, in works that are subject to higher mean temperatures, the resistance should be corrected by the temperature effect.

In the absence of experimental values, long-term resistance can be extrapolated from short-term resistance (T_r) obtained from the wide-width tensile strength test, and applying the factor for reduction due to creep (f_f):

$$T_k(t) = \frac{T_r}{f_f}$$

To evaluate the reduction factors in the absence of certified experimental data, Tables 3.5 and 3.8 can be used. Some types of reinforcements are not included in the tables because of lack of experimental information and manufacturers should be consulted directly (this is the case of polyaramid geogrids, for example).

The reduction factors due to environmental degradation must be carefully evaluated, discarding the use of any given geosynthetic if excessive degradation is suspected.

3.9 STRESS INCREMENTS DUE TO SURCHARGE LOADINGS

Surcharges applied to the top of the reinforced structures promote increased internal stress in the reinforced soil structures. Reinforcements should be designed to withstand the summation of the tensile loads due to the wall's own weight, to surcharge loading and to constructive efforts (compaction). Soil compaction has the advantage of minimizing post-constructive strains. When surcharge loading are applied, significant strains occur only for values greater than the vertical stress induced by compaction.

The horizontal stress increment due to compaction is "memorized" by the soil and can generate significant increases in tensile strengths on reinforcements, as discussed before. However, these increases tend to dissipate partially over time, through tensile relaxation of the geosynthetics and the soil itself. The speed of the process depends on the characteristics of the materials involved, i.e., soil and the polymer corresponding to the adopted reinforcement.

TABLE 3.5 CREEP FACTORS FOR SEVERAL POLYMERS - f_f
(adapted from Azambuja, 1999)

Polymer	Creep factor (f_f)
Polyester (PET) and PVA	2.0 - 2.5
Polypropylene (PP)	3.5 - 4.0
High density polyethylene (HDPE)*	4.5 - 5.0

* Valid for work hardening oriented polymers

TABLE 3.6 CRITERIA FOR DETERMINING THE SEVERITY OF THE ENVIRONMENT (Allen, 1991)

Type of equipment	Soil	layer thickness		
		< 15 cm	15 to 30 cm	> 30 cm
Light and towed	Fine to coarse sand, Sub-rounded grains	low	low	low
	Sand and graded gravel, sub angular grains, $\varnothing$< 75 mm	moderate	low	low
	Poorly graded gravel, angular grain, $\varnothing$> 75 mm	very high	high	moderate
Self-propelled	Fine to coarse sand, Sub-rounded grains	moderate	low	low
	Sand and graded gravel, sub angular grains, $\varnothing$< 75 mm	high	moderate	low
	Poorly graded gravel, angular grain, $\varnothing$> 75 mm	not recommended	very high	high

When there is a long enough time gap between the end of construction and the application of the overloads in certain situations, significant reductions in efforts in the soil and on reinforcements could occur.

Theoretically, it would be possible to estimate this stress reduction, provided that information about the time × stress × strain behavior of the geosynthetics and the soil were available. In the absence of such information, the structure must be designed conservatively in order to support the summation of efforts, in the calculations of both internal and external stability.

TABLE 3.7 MECHANICAL DAMAGE REDUCTION FACTORS - f_d (Azambuja, 1999)

Geosynthetic	survivability	severity of the environment			
		low	moderate	high	very high
PP woven Geotextile	low	1.30-1.45	1.40-2.00	NR	NR
	moderate	1.20-1.35	1.30-1.80	NR	NR
	high	1.10-1.30	1.20-1.70	1.60-NR	NR
PET woven Geotextile	high	1.10-1.40	1.20-1.70	1.50-NR	NR
	low	1.15-1.40	1.25-1.70	NR	NR
Nonwoven PET or PP geotextile	moderate	1.10-1.40	1.20-1.50	NR	NR
	high	1.05-1.20	1.10-1.40	1.35-1.85	NR
Flexible Geogrid . acrylic coating	moderate	1.10-1.20	1.20-1.40	NR	NR
	high	1.10-1.15	1.20-1.40	1.50-NR	NR
Flexible Geogrid. PVC coating	moderate	1.05-1.15	1.15-1.30	1.40-1.60	NR
	high	1.05-1.15	1.15-1.30	1.40-1.60	1.50-2.00
PP rigid geogrid	moderate	1.05-1.15	1.05-1.20	1.30-1.45	NR
PEAD rigid Geogrid	moderate	1.05-1.15	1.10-1.40	1.20-1.50	1.30-1.60
	high	1.04-1.10	1.05-1.20	1.15-1.45	1.30-1.50

Note: NR - not recommended (loss of resistance exceeding 50%)

TABLE 3.8 BIOCHEMICAL DEGRADATION FACTORS - f_a, FOR NON CERTIFIED GEOSYNTHETICS (AZAMBUJA, 1999)

Reinforcement function	Polymer	reduction factors (f_a)
Temporary	Polypropylene	1 – 1.25
	Polyethylene and PVA	1 – 1.25
	Polyester	1 – 1.25
Permanent	Polypropylene	1.25 – 2.0
	Polyethylene and PVA	1.1 – 1.5
	Polyester	1.25 – 2.0

Note: highest values refer to severely unfavorable environments - strongly acidic for Polyolefin and strongly alkaline for polyester.

To estimate the tensile increment due to surcharge loading applied on the embankment, limit equilibrium methods or methods based on the theory of elasticity are often used, despite the theoretical contradictions arising, according to Palmeira and Gomes (1996).

3.9.1 Stress increments resulting from uniform surcharge loadings

In the case of uniform infinite loads, analysis of the corresponding tension increments considering the external load in the calculation of the vertical geostatic stress could be performed, as shown in Sections 3.4.3.

For an uniform load over an area, Mitchell and Villet (1987) present the procedure described below.

The tensile stress increments at reinforcement due to application of external load, as shown in Figure 3.10, and given by:

$$\Delta T_{max} = K \cdot \Delta\sigma_z \cdot S_v$$

where ΔT_{max} is the additional maximum tensile strength at reinforcement; K is the pressure coefficient; $\Delta\sigma_z$ is the additional vertical stress due to load, where T_{max} operates; and S_v is the vertical spacing.

The method considers that K decreases linearly with depth,

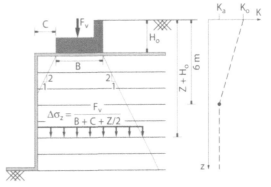

Figure 3.10 *Vertical stress increments and earth pressure coefficient versus depth, according to Mitchell and Villet (1987)*

from the surface where $K = K_0$, up to 6 meters deep, where $K = K_a$. For greater depths, $K = K_a$ is considered. The procedure suggested by the authors is simplistic and does not consider the influence of important factors in the maximum stresses at the reinforcement, such as soil compaction and relative soil-reinforcement stiffness. The simple consideration of a linear variation in the K value along the depth may be incorrect, especially when the compaction effort is high.

To consider the stiffness of the reinforcement and compaction for the calculation of the tensile stress in the reinforcement, a procedure adapted from the methods of Mitchell and Villet (1987), and Ehrlich and Mitchell (1994) can be used. In the calculation of the vertical geostatic stress (σ'_z), one must consider the increase from the external load: ($\Delta\sigma'_z$), as is suggested in Mitchell and Villet (1987):

$$\sigma'_z = \gamma \cdot z + \Delta\sigma'_z$$

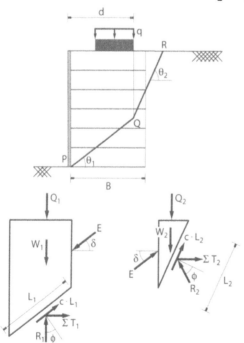

From this, the stress at the reinforcement is calculated as shown in Sections 3.4.3. Andrade, Ehrlich and Iturri (1999) used numerical analysis to show that this procedure provides consistent results.

3.9.2 Other solutions for stress increments

Palmeira and Gomes (1996) conducted studies in reduced models and concluded that the methods based on the theory of elasticity provides satisfactory results for values of $d/B \approx$ 0.5, as shown in Figure 3.11. For loads further from the face of the structure, the authors state that the solutions based on the theory of elasticity may underestimate the horizontal stress increments. In this case, the authors

Figure 3.11 *Bilinear critical surface method for horizontal stress increments (adapted from Palmeira and Gomes, 1996)*

suggest the method of critical bilinear surface, also shown in Figure 3.11.

This method uses the equilibrium of horizontal forces, vertical forces and moments of wedges. The unknown quantities of wedge 1 are the reaction of the stable soil (R_1), the thrust between wedges (E) and the summation of the required tensile forces in the reinforcements intercepted by the L_1 segment (ΣT_1). The inclination of the E force is arbitrated, ranging from zero (conservative) to ϕ'. W_1 and W_2 are the weights of each wedge. Q_1 and Q_2 are the resultants of loading on the upper surface of the wedges. After solving the system of equations for wedge 1, the system for wedge 2 is solved using the value of E obtained for wedge 1. The location of points P, Q and R is unknown *a priori*, so the method requires further search for the most critical surface. The safety factor of for each wedge is defined by:

$$FS = \frac{\sum T_{d,i}}{\sum T_i}$$

where $\Sigma T_{d,i}$ is the summation of the reinforcement design resistances for wedge "i".

It is noteworthy that ΣT_i should also consider the efforts resulting from compaction.

Besides the solutions presented, there are several others, usually based on limit equilibrium methods, with the disadvantages cited before. Other methods, such as those based on finite element programs can also be used.

3.10 LATERAL DISPLACEMENT AND STRAIN PREDICTION

Several researchers have studied the horizontal displacements that occur during the construction of reinforced soil structures.

Elias, Christopher and Berg (2001) present a curve that allows for the empirical estimation of the value for maximum horizontal displacement (u_{max}) due to construction in reinforced soil structures. The procedure, under the auspices of the FHWA, is based on measurements taken from walls up to 6 meters in height; and allows the estimation of u_{max} for extensible or inextensible reinforcements, according to the height of the wall and the length of the reinforcements. The authors also suggest an increase to the u_{max} value by 25% for each 20 kPa of surcharge, but they acknowledge that actual horizontal displacements of the wall will be influenced by soil type and compaction energy. Figure 3.12 shows the proposed ratio.

Although the method is based on reinforced soil structures that were constructed in countries with temperate climate, Becker (2006) applied it

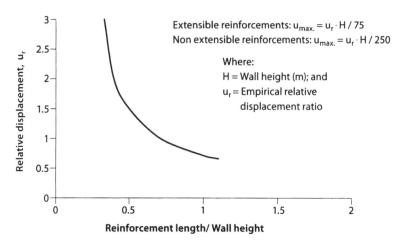

Figure 3.12 *Empirical curve to estimate the maximum horizontal displacements of reinforced soil walls, according to Elias, Christopher and Berg (2001)*

to a Brazilian structure. The author monitored the construction of a wall with residual clayey silt soil, reinforced with PVA geogrids, 5-meters high and slope face 5V: 1H, noting a good agreement between the measured horizontal shifts and the estimate provided by the method.

It is important to note that the face slope, the characteristics of the foundation, soil properties and reinforcement spacing are not taken into account. However, the method is simple to use. For example, for a 4.5 meter high wall, with extensible reinforcements with length equal to 93% of the height, we obtain an estimated $u_{max} = 45$ mm.

The movements in reinforced soil masses are originated by several causes: settlements at the base of RSS, rotations per eccentricity of loads, distortion and sliding of reinforced soil mass resulting from the earth pressure in the non-reinforced area and strains caused by tensile stresses to which they are submitted. Ehrlich (1995) presented an analysis of the movements of an instrumented reinforced soil wall and concluded that the most relevant portion of the observed movements result from strains on reinforcements. This conclusion remains valid for conventional RSS built on competent soil foundations.

Following are procedures for estimating the movements caused by the strains on reinforcements. To estimate the remaining portions of the RSS movements, it is recommended that procedures described in Ehrlich (1995) be used.

Consider a reinforcement with J_r stiffness subjected to an idealized distribution of loads along its length, as shown in Figure 3.13.

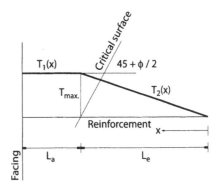

The maximum reinforcement tension T_{max} occurs at the point of intersection between the reinforcement and the critical surface. The tension distribution functions along portions of the reinforcement that are embedded in the active and resistant zones are, respectively, $T_1(x)$ and $T_2(x)$. The distribution functions of strains $\varepsilon_1(x)$ and $\varepsilon_2(x)$ in the same areas can be written as follows:

Figure 3.13 *Idealized distributions of tensile loads along reinforcement*

$$\varepsilon_1(x) = \frac{T_1(x)}{J_r} = \frac{T_{max}}{J_r}$$

$$\varepsilon_2(x) = \frac{T_2(x)}{J_r} = \frac{T_{max} \cdot x}{L_e \cdot J_r}$$

The total horizontal displacement u suffered by the reinforcemnt at the face:

$$u = \int_0^{L_e} \varepsilon_2(x)\,dx + \int_{L_e}^{L_e+L_a} \varepsilon_1(x)\,dx$$

$$= \frac{T_{max}}{J_r} \cdot \left[\int_0^{L_e} \frac{x}{L_e}\,dx + \int_{L_e}^{L_e+L_a} dx \right]$$

$$= \frac{T_{max}}{J_r} \cdot \left[\left. \frac{x^2}{2 \cdot L_e} \right|_0^{L_e} + |x|_{L_e}^{L_e+L_a} \right]$$

$$= \frac{T_{max}}{J_r} \cdot \left[\frac{L_e}{2} + L_a \right]$$

For reinforcements with length $L_r = L_e + L_a$, at height z, the horizontal displacement along the surface can be expressed as follows:

$$u = \frac{T_{max} \cdot L_r}{2 \cdot J_r} \cdot \left[1 + \frac{z \cdot \tan(45 - \phi'/2)}{L_r} \right]$$

Thus, one can determine the horizontal displacement of each reinforcement based on the maximum tensile force. With the horizontal displacements in all the layers, one can draw the strained shape of the face, taking into account the constructive process that was employed.

The J_r stiffness modulus of the reinforcement decreases over time due to tensile relaxation in soil-geosynthetic assembly. For the same reason, the ground stresses should also decrease, resulting in a reduction of the maximum acting tensile force T_{max}. The analysis of this combination of effects is very complex. However, the effects tend to override each other; it is suggested that the J_r value obtained with wide-width tests be adopted for the determination of the maximum reinforcement tension and for the estimation of facing displacements. It bears repeating that the determined T_{max} value must not exceed, under any circumstances, the project tensile strength T_d available at the end of the structure's useful life.

Generally, the facing location is checked only in the first layer of the RSS. Each of the following layers is positioned according to the previous layer. Thus, the displacements accumulate as the wall is built. Ehrlich (1995) reports the case of reinforced soil wall with wrap around geotextiles with designed face slope 1H: 8V. Because of the accumulated displacements at the end of construction, the wall presented a vertical face. Whenever this constructive procedure is adopted, the final horizontal displacement of each reinforcement layer will be the sum of the value calculated for the layer itself plus the values calculated for the lower layers.

Becker (2006) monitored the horizontal displacements of a reinforced soil wall with wrap around geogrids that was built differently. To reduce the deviations from the design, it was decided that each of the reinforcement layers would be placed individually by topography. This way, the displacements suffered earlier by the lower layers did not affect the overall displacement of the upper layers. When this constructive procedure is adopted, the displacements calculated for the lower layers should not be accumulated. The displacement for each layer depends only on the tensile strength acting upon it.

Figure 3.14 shows typical displacement profiles for the two positioning methods of facing. The drawing is not set to scale for better comprehension.

It should be noted that the method presented is approximate; since it considers that the compaction of the upper

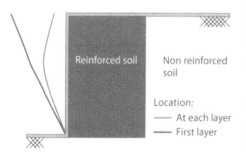

Figure 3.14 *Displacement profiles depending on the positioning method of facing (not to scale)*

layers does not influence the strains and the tensile stresses that occur in a given layer.

The compaction effort applied to the upper layers generate tension increases in the lower layers, which, as noted earlier, do not outweigh the increase generated by compaction on to the layer. Thus, the residual horizontal tensile strength that remains in the reinforcement depends only on the compaction of the layer where it was placed, as assumed by the default method of Ehrlich and Mitchell (1994). However, in the case of strains, since the soil is not an elastic material, each stress increase leads to an additional strain, not considered by the method. Fortunately, the largest share of strains suffered by a reinforcement refers to the compaction of the layer itself. Field measurements (Becker, 2006) indicated that approximately 70% of the strains measured by telltales installed in the geogrids of a RSS result from the compaction of the layer where the reinforcement is placed.

Another alternative for strain estimation is the use of finite element software.

Finally, it is important to understand that the use of high-energy compaction equipment (which lead to greater constructive strains) is not detrimental to the performance of the structure. On the contrary, when applying a low compaction effort, there are smalller constructive strains, but the compacted soil obtained is more deformable. Thus, the loads applied to the retaining structure during its service life, would result on greater post-constructive strains. When high compaction efforts are employed, we obtain structures that are much less susceptible to post-constructive strains, caused by loads applied during service life. In this sense, one can compare the compaction of a reinforced wall with the pre-stressing of a tieback wall. The larger the pre-stressing loads or compaction efforts, the smaller the post-constructive displacements.

Evidently, the stiffness and strength of reinforcements must be compatible with the compaction effort applied in the same way that tiebacks should have compatible resistance with the pre-stressing effort.

3.11 CASE STUDIES

Among the various walls and slopes constructed in Brazil with residual or lateritic soils, some were instrumented for performance monitoring. The observed results were always favorable. For more details, it is recommended to consult the works cited in Table 3.9.

Table 3.9 Characteristics of some instrumented RSRS in Brazil

Reference	Height (m)	Width (m)	Face	Face inclination
Carvalho, Pedrosa and Wolle (1986)	10.0	7.0	Wrap around	1H:2V
Other characteristics	▶ Spacing: 0.6 m ▶ Type of reinforcement: woven and nonwoven geotextiles ▶ Tensile strength: 22 kN/m ▶ Soil type: Clayey sand with silt (residual granite)			
Ehrlich, Vianna and Fusaro (1994)	4.0	2.8	Wrap around	1H:8V
Other characteristics	▶ Spacing: 0.3 m ▶ Type of reinforcement: nonwoven geotextile ▶ Tensile strength: 22 kN/m ▶ Soil type: residual sandy clay with silt ▶ compaction equip: hand made			
Bruno and Ehrlich (1997)	3.9	3.5	Wrap around	1H:8V
Other characteristics	▶ Spacing: 0.6 m ▶ Type of reinforcement: nonwoven geotextile ▶ Tensile strength: 22 kN/m ▶ Soil type: residual sandy clay with silt ▶ compaction equip: hand made			
Becker (2006)	5.0	4.2	Wrap around	1H:5V
Other characteristics	▶ Spacing: variable 0.4 to 0.6 m ▶ Type of reinforcement: PVA geogrid ▶ Tensile strength: 55 and 35 kN/m ▶ Soil type: residual silty clay ▶ compaction equip: Rammer Compactor and Dynapac CA25 Roller			
Benjamin (2006)*	4.0	3.0	Wrap around	1H:5V
Other characteristics	▶ Spacing: 0.4 m ▶ Type of reinforcement: PP woven, PET and PP nonwoven geotextiles ▶ Tensile strength: variable 8,83 kN/m to 54,24 kN/m ▶ Soil type: residual silty clay and silty sand (with clay) ▶ compaction equip: Wacker BPS 1135 W vibrating plate			
Riccio (2007)	4.2	3.0	Segmental blocks	1H:10V
Other characteristics	▶ Spacing: variable 0.4 to 0.6m ▶ Type of reinforcement: PET geogrid ▶ Tensile strength: 55 and 35 kN/m ▶ Soil type: residual sandy clay ▶ compaction equip: Rammer Compactor and Dynapac CA25 Roller			

* Prototype walls built for research purposes

CONSTRUCTIVE ASPECTS 4

4.1 TYPE OF REINFORCEMENT

There are various types of geosynthetics available to be used as reinforcement in retaining systems: rigid and flexible geogrids, woven, nonwoven and reinforced geotextiles; geocells, geobars, strips and the use of diffuse fibers (micro-reinforcement). To date, geogrids and geotextiles have a more significant role as reinforcements for the construction of walls and reinforcement of steep slopes, which is why the discussion about the type of reinforcement will be focused on these geosynthetics.

Besides the structure itself, there are several types of polymers available, and the most frequently used are: polyester, polypropylene and high density polyethylene. Recently, fiber polyaramid has been used for special reinforcements that require particularly high tenacity (Ehrlich, Azambuja, 2003).

The choice of a specific type of reinforcement for a reinforced soil retaining system design requires an economic performance analysis, since any product can be used, provided that the mechanical and rheological properties are known. However, there are some conditions or structural features of the environment that favor certain sets of geosynthetics and therefore should be observed when choosing the type of product or polymer to be used.

4.1.1 Height of wall

The greater the height of the wall, the greater the strength required from reinforcements. In general, geogrids are more favorable for walls higher than 4 meters, while geotextiles often present economic advantages for smaller walls (Ehrlich, Azambuja, 2003).

4.1.2 Strain/Deformations Restrictions

In walls where the dimensional tolerances are important constraints or post-construction strains are undesirable, geogrids are most appropriate, followed by reinforced geotextiles. The strains due to creep are more diffi-cult to predict in polyolefin reinforcements (high-density polyethylene and polypropylene) and impose restrictions on products with such polymers when post-construction strains (long term) are critical constraints. In such cases, the use of polyester or polyaramid reinforcements usually presents economic advantages.

4.1.3 Severity of the environment

The chemical aggressiveness of the soil or of the environment where the retaining system shall be deployed is another important factor to consider for product choice. In chemically adverse environments, geogrids are less sensitive than geotextiles. This is due to greater thickness of its elements, which reduces the surface area of exposure, or to the protective coating that some products have. Certain acidic environments, for example, are restrictions on the use of polyester based geosynthetics , due to degradation of these polymers by hydrolysis (Ehrlich, Azambuja, 2003).

In addition to the chemical aggression, we must consider the severity of the environment towards the mechanical damages of installation. Coarse soils, with angular grains damage woven geotextiles more profoundly than nonwoven ones. Rigid geogrids present lesser damages than flexible ge-ogrids. These sensitivities should be considered in determining permissible resistances of reinforcements or in the prescription of the soil type for each specific project situation, as mentioned in section 3.8.

4.2 TYPE OF FACING AND CONSTRUCTION PROCEDURE

In the first reinforced soil structures with geosynthetics, a constructive technique called wrapped facing was used. Later, constructive systems evolved to use other forms of facing. The types of systems most commonly used in practice (Ehrlich, Azambuja, 2003) will be discussed below.

4.2.1 Wrapped facing systems

The technique is comprised of the conformation of systems where the geosynthetic confines the soil laterally between two layers of reinforcement, by means of its fold and anchoring inside the wall. In most cases, the wall or reinforcement of steep slopes is erected with the help of lateral and removable light elements, and then a permanent face protection system is built. Self-enveloping systems can be built with the anchoring at the base (Fig. 4.1) or at the top layer (Figure 4.2). Typically, these anchors have a minimum length of 1 meter, but may be longer if the design requires it. The final facing can be constructed with various techniques, from masonry walls to shotcrete. The techniques with best success rates are those where

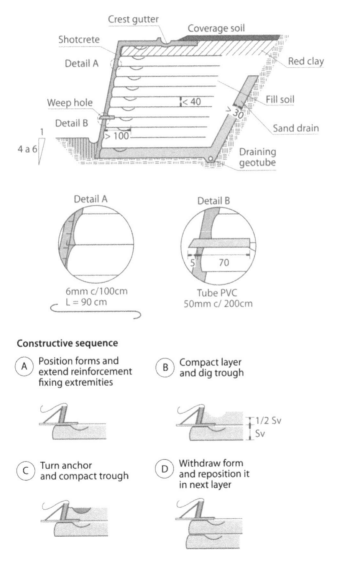

Figure 4.1 *Typical cross section of wall with wrapped facing systems with top anchor (adapted from Ehrlich and Azambuja, 2003)*

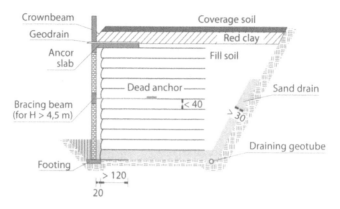

Figure 4.2 *Typical cross section of wall with wrapped facing systems with lower anchor (adapted from Ehrlich and Azambuja, 2003)*

the wall is built slightly away from the reinforced soil structure, which reduces the post-construction strain effects on the aesthetics of the wall.

Other attractive construction procedures for a wrapped facing wall are the ones that use a lost-form, consisting of an electro welded wire mesh and metallic anchors. These techniques are better suited for the use of geogrids, and the final facing can be constructed with shotcrete (Figure 4.3) for walls or with ground cover for less steep facings (Ehrlich and Azambuja, 2003).

4.2.2 Systems with segmental blocks

The system with segmental blocks consists of the use of prefabricated concrete elements, which are used as a lateral compaction mold for the layers, while they are also the final facing. Most of these systems are composed of light blocks that can be assembled manually by a worker.

The blocks have fitting devices between them, so that the alignment of the wall is facilitated during construction. At the same time, it provides an efficient anchor for reinforcements. Because of this characteristic, segmental blocks are also called "interlocking blocks".

There are many such constructive systems with industrial properties, some of them adapted for a specific type of reinforcement - in most cases, geogrids. An example of a segmental block system is presented in Figure 4.4.

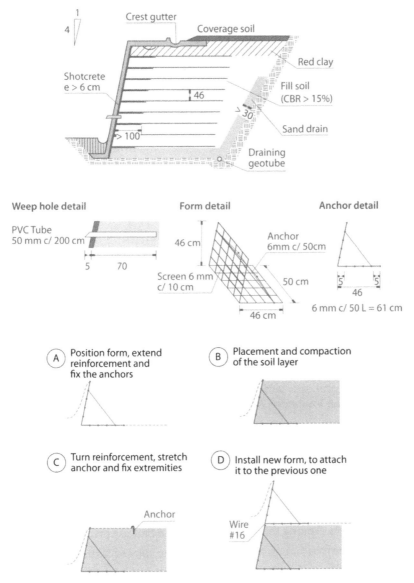

Figure 4.3 *Typical cross section of wrapped facing wall and lost-form construction procedure (Ehrlich and Azambuja, 2003)*

4.2.3 **Hybrid systems**

Some systems with segmental blocks are associated with wrapped facing techniques, constituting a hybrid method, mostly used for reinforcements with geotextiles. These are techniques in which the final facing is also

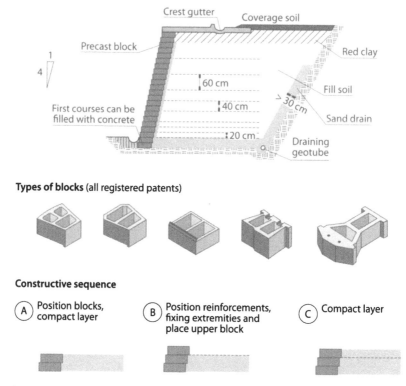

Figure 4.4 *Typical cross section of wall with segmental block system (adapted from Ehrlich and Azambuja, 2003)*

used as a mold, but the reinforcement is not connected to the facing. This implies difficulties in the maintenance of vertical and horizontal alignment of the structure.

Generally, the blocks are anchored to the mass by secondary reinforcements. Among these systems are those that use precast concrete "L" shaped elements as facing, which present some constructive entanglements due to excessive weight and less flexibility of the elements (Ehrlich and Azambuja, 2003).

4.2.4 Systems with modular panels and integral panels

The construction processes of reinforced soil with modular panels was disseminated with the wide employment of reinforced earth. In such cases, the facing also acts as a mold; however, the connection of the panels with the reinforcements and between the panels is more complex. Although relatively slim, modular panels are heavy and require mechanization for handling.

Integral walls, in turn, are tall structures in which each element of the face is the total height of the wall. These systems are applied to walls that require special aesthetical features of the facing, since there is great difficulty to manage the parts.

Both the modular panels and the integral walls are systems better suited for less extensible reinforcements, such as geobars or straps with high tenacity polymers, since they do not tolerate significant constructive deformations/strains (Ehrlich and Azambuja, 2003).

4.2.5 Comparison between constructive systems

Among the various constructive systems available, the designer or developer will have to choose the technique that is most appropriate for each case, considering the cost, the mechanical function and architectural function of the retaining structure. However, it should be noted that despite the effectiveness of all techniques, the cost and aesthetic aspects vary significantly.

Wrap around walls are usually built with geotextiles, but geogrids can also be used, provided there is a system to prevent the escape of soil through the facing system (lost-form system). With this system, one can achieve a lower cost, but dimensional control is difficult and facing appearance falls short of expectations. These systems also have a higher tolerance to settlements, because of the low stiffness of the facing.

Segmental block systems usually have good aesthetic finish and are often used in works where facing is apparent. These systems have a reasonable tolerance to settlements and allow for easy dimensional control, which is why they are best suited for builders with no previous experience in reinforced soil works.

Modular panel or integral wall systems should not be built with low stiffness reinforcements. The facing appearance is good, as is dimensional control, however, costs tend to be higher, and the tolerance to settlements is lower.

4.3 ARRANGEMENT OF REINFORCEMENTS

Reinforcements may be arranged with uniform or variable vertical spacing. In wrapped facing systems, uniform spaces are recommended due to constructive issues, especially for control of constructive strains, although such practice leads to higher consumption of geosynthetics. With segmental block systems, variable spacing between layers allows a better rationalization of the system, but of course, the maximum spacing between reinforcements depends on the capacity of the blocks to absorb the stresses generated by the compaction without rupturing a portion of the face.

Instability problems on the face due to large spacing can be solved with the use of short secondary reinforcements between the layers of the main reinforcements, in any constructive system.

In general, limiting the vertical spacing (S_v) to 0.80 meter is recommended. For segmental blocks systems, it is recommended that the spacing not be greater than twice the depth of the blocks.

4.4 DRAINAGE SYSTEM

Draining is one of the most important aspects of building reinforced soil walls. In all currently used design methods, it is assumed that pore pressure is null in the reinforced soil mass.

Tropical soils, usually compacted near the optimum moisture content, are partly saturated and usually have considerable amounts of suction (negative pore pressure due to partial saturation). Suction is difficult to estimate and it is not considered in the calculations, but has the effect of increasing stability.

The presence of water level within the reinforced soil is doubly undesirable because it leads to positive pore pressure, invalidating the hypothesis of null pore pressure and it eliminates suction. In both cases, the effect is reduced structure safety.

To prevent this from happening, the reinforced soil must be provided with adequate draining systems.

It is strongly recommend to employ a draining blanket with thickness between 20 and 50 cm under the reinforced soil mass and between the mass and the non-reinforced area.

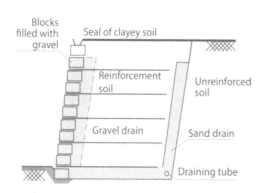

Figure 4.5 *Draining blanket system*

Figure 4.5 shows a typical cross section of a reinforced soil wall with segmental block facing with a draining blanket between the reinforced soil mass and the natural soil. It is also evident that another layer of draining material is placed near the face and the blocks are filled with gravel. The top of the structure is covered by a layer of compacted clay material to prevent seeping of rainwater, and it is advisable to install drain gutters. The draining material may be composed of clean sand or pebbles. To play the role of a filter and prevent the escape of fines from the reinforced and non-reinforced soil, the draining material can be encased in geotextiles or grain size of the material can be selected so that the granulometric curve is not discontinuous and it meets the following filtration criteria (Terzaghi; Peck, 1967):

$$5 \cdot D_{15\,soil} < D_{15\,drain} < 5 \cdot D_{85\,soil}$$

where $D_{15\,soil}$ is the diameter below which are 15% of soil weight; $D_{15\,drain}$ is the diameter below which are 15% of draining material; and $D_{85\,soil}$ is the diameter below which are 85% of soil weight.

Besides the drainage of groundwater, there should be a superficial draining/waterproofing system to prevent erosion and infiltration of rainwater into the reinforced soil mass, which usually consists of longitudinal and transverse gutters (stairways), catch basins, storm drains, etc.

4.5 BACKFILL MATERIAL

Because of transport costs, one can say that, except for useless materials, the best soil quarry is the one that is nearest. It is recommended that the first soil search for the work be done through be the observation of cut slopes or excavations near the work site and auger surveys. The soil thus obtained should be classified by visual-tactile analysis carried out by experienced staff.

Obviously useless soils should be discarded immediately, such as peat, organic clays, soft clays, granular soils rich in mica etc.

After verifying that the chosen soil is enough in quantity for the requirements of the work, a representative sample of about 20 kg and 40 kg, in case of fine or coarse soil, respectively, should be sent to the laboratory for testing as listed in Table 4.1. The characterization test results can be used to classify the soil according, for example, to the Unified Soil

QUADRO 4.1 SOIL CHARACTERIZATION TESTS

Test	Brazilian Standards	Notes
Atterberg Limits	NBR 7180/84 Soil – Determination of plastic limit NBR 6459/84 Soil – Determination of liquid limit	Except for soils without fine fraction
Grain size	NBR 7181/84 Soil – Grain size analysis	With dispersant
	NBR 13602/96 Soil – Dispersive characteristics of clay soils by double hydrometer – Method of test	With and without dispersant
Density of grains	NBR 6508/84 Soil grains finer than 4.8 mm – Determination of density	
Natural water content	NBR 6457/86 Soil samples – Preparation for compaction and characterization tests	Laboratory
	DNER-ME 52/94 Soils and small aggregates – Determination of water content through the use of "Speedy"	Field
	DNER-ME 28/61 Quick stove-top method for determining the water content of soils	Field
	DNER-ME 88/64 Quick alcohol method for determining water content of soils	Field
Proctor	NBR 7182/86 Soil – Compaction test	Except for soils without fine fraction

Classification System described in Table 2.2. However, it is known that this classification is inappropriate for tropical soils, whether lateritic or saprolite, since these soils are a product of very different evolutionary processes from those in temperate climates, and they behave differently. For example, high percentages of fine particles in soils of colder climates usually lead to poor mechanic performance, which does not occur in tropical soils.

The criteria for acceptance of a specific soil, therefore, must be based on the experience of local use of that soil or on mechanical tests (resistance and deformability) with compacted samples. It should be noted that the MCT classification, developed in Brazil by Nogami and

Villibor (1981, 1995), with the main purpose of identifying, classifying and estimating the performance of tropical soils for roadwork purposes, may be advantageous in estimating the behavior of soils. In this classification, the soils are identified based on their properties after compaction, according to standard procedures and testing of mass loss by immersion, and classified into two classes (lateritic and non-lateritic behavior) and seven groups, which depend on grain size/grading of the soil. The MCT classification grants a performance evaluation for each group, according to expansion, contraction and permeability coefficient properties, and it assesses the suitability of each group to be used as fill, sub grade, pavement base etc. More details about this method can be found in Cozzolino and Nogami (1993).

Most international prescriptions recommend the use of granular non-plastic soils (PI <4%) or impose restrictions to the percentage of fines. These rules aim to avoid the clayey soils of temperate countries, which have admittedly unfavorable performance. For countries with tropical soils, however, the use of lateritic soils is particularly advantageous because they have significant cohesion, without a tendency of excessive plastification or creep. Thus, for the Brazilian reality, it is recommended that the plasticity index be less than 20%, although there are known successful cases that used soils with plasticity index of up to 30%.

Azambuja and Strauss (1999) reported eleven cases of reinforced soil structures built in the state of Rio Grande do Sul, with heights of up to 9.5 meters, resulting from their professional experience. In nine reported cases, the soils are described as clayey or residual, the latter derived from rocks such as granite, basalt and even argillite.

For resistance control purposes and also to control the degradability of the material, it is recommended that the support index (CBR) be greater than 15% and the expansion by saturation at optimum water content content be less than 2% (Ehrlich and Azambuja, 2003).

In general, it can be said that any soil that is suitable for compaction of non-reinforced fills can lend themselves to the construction of reinforced soil structures, provided that the necessary draining measures are taken, as discussed in section 3.4.

The compaction of reinforced soil walls should be managed taking into account the soil type (severity of the environment), the type of reinforcement (survivability) and the desired resistance in the reinforced area. It is convenient, within reason, that the placing and compaction work be mechanized. Due to face sensitivity, it is highly recommended that compaction be carried out with lightweight equipment - for example, rammer compactors, ranging from 0.5 meter to 1 meter in width, adjacent to the facing. The use of vibrating plates in clayey soils is not recommended, as they result in poor compaction.

Compaction control should be performed at each layer, for volumes less than 300m³, or according to the scale of the work. As it is difficult to correct water content in reinforced structures, it is recommended to adjust it at the bed/quarry or on a stretch specified for this purpose before compacting in order to prevent damages to the geosynthetics during plough operations.

4.6 QUALITY CONTROL AND CONSTRUCTIVE TOLERANCES

According to Ehrlich and Azambuja (2003), three aspects are important in quality control of reinforced soil retaining walls: nominal resistance of the reinforcements, mechanical damage control and control of strains/deformations during construction. It is recommended that the reinforcements be tested per batch and at each 1,000 m² of reinforcements. The minimum recommended tests for reinforcement control are wide width strength and puncture, and they should provide results compatible with the nominal strengths, for a 95% confidence level.

Constructive strains/deformations should be controlled at each layer. The distortions of the face (ratio between horizontal displacements at the crest and the height of the wall) should be less than 1% for integral walls and panels, 2% for segmental blocks and 5% for wrap around systems (before the final face).

4.7 GUARDRAILS

The designer should be careful with vehicle impact problems at the top of the reinforced soil walls, where they usually hold roadbeds. Guardrails should not be directly attached to the facing, but to an independent retaining structure (Ehrlich and Azambuja, 2003).

Design Example

5.1 Soil and wall characteristics

Consider the design of a reinforced soil wall with interlocking concrete blocks facing and geogrids, with the following characteristics. For the internal design, the abacus proposed by Ehrlich and Mitchell (1994) will be used.

Geometry of the wall

Height of wall	H = 5.8 m
Spacing of reinforcements	Sv = 0.60 m (constant)
Face slope	1H:10V

Parameters of fill soil:

Description	clayey silt lateritic soil, PI = 15%
Unit weight	γ = 18 kN/m^3
Angle of friction	ϕ' = 35°
Cohesion intercept	c' = 10kPa
Initial tangent modulus	K = 128
Exponent modulus	n = 0.78

Parameters of the foundation soil:

Same as fill soil

Parameters for blocks and reinforcements:

Type of reinforcement	PVA geogrid
Coupling Efficiency	85%
Block dimensions	40 cm x 40 cm and height of 20 cm

Compactor roller features:

Type of equipment	Dynapac CA134PD self-propelled roller
Induced vertical stress	110 kPa (Fig. 2.18)

5.2 EXTERNAL STABILITY ANALYSIS

5.2.1 Determining the length of reinforcement

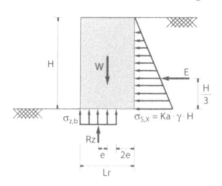

Figure 5.1 *Forces and variables involved in the external stability analysis (Sayao et al., 2004)*

The length of the reinforcements (L_r) should be defined as to ensure external stability. Figure 5.1 shows the forces and variables involved in the analysis. For simplicity of the determinations, the calculations will be conducted considering the vertical face of the wall.

a) Thrust of the unreinforced area

The earth thrust (E), in the absence of surcharge actions, is given by:

$$E = \frac{1}{2} \cdot \gamma \cdot H^2 \cdot Ka$$

The earth pressure coefficient in an active state (K_a), considering the horizontal embankment and neglecting the friction between the soil and the wall, in favor of safety, can be calculated by the formulation of Rankine:

$$K_a = \tan^2\left(45° - \frac{\phi}{2}\right) \quad \Rightarrow \quad K_a = \tan^2\left(45° - \frac{35°}{2}\right) = 0.27$$

b) Slide verification

The sliding safety factor (FS) is determined by the ratio between the shear resistance at the base of the wall and the earth pressure:

$$FS = \frac{\gamma \cdot H \cdot L_r \cdot \tan\phi'}{\frac{1}{2}\gamma \cdot H^2 \cdot Ka}$$

Once the minimum safety factor of 1.5 is established, the length of the reinforcement (L_r) for a wall without surcharge loading action is given by:

$$L_r \geqslant \frac{3}{4} \cdot \frac{H \cdot Ka}{\tan\phi'} \quad \Rightarrow \quad L_r \geqslant \frac{3 \times 5.8 \times 0.27}{4\tan 35°} = 1.68 \, m$$

c) Overturning verification

The overturning safety factor (FS) is defined by the ratio between the stabilizing moments provided by the weight of the wall and the

unstabilizing moment generated by the horizontal thrust of the soil (E),
so that:

$$FS = \frac{W \cdot \frac{L_r}{2}}{E \cdot \frac{H}{3}} = \frac{\frac{\gamma \cdot H \cdot L_r^2}{2}}{\frac{Ka \cdot \gamma \cdot H^3}{6}} = 3 \cdot \frac{L_r^2}{Ka \cdot H^2}$$

Once the minimum safety factor of 2.0 is established, the length
of the reinforcement (L_r) for a wall without surcharge loading action is
given by:

$$L_r \geqslant \sqrt{\frac{2}{3} \cdot Ka \cdot H^2} \quad \Rightarrow \quad L_r \geqslant \sqrt{\frac{2 \times 0.27 \times 5.8^2}{3}} = 2.47\,m$$

d) Verification of stresses at the base

The resultant from stresses acting at the base of the wall must
guarantee that the whole base is under compression and, therefore, the
eccentricity (e) must be less than a sixth of L_r.

The eccentricity of the resultant of the normal stresses acting at the
base of the wall (e) can be expressed as:

$$e = \frac{E \cdot \frac{H}{3}}{Rz} = \frac{\frac{1}{2} \cdot \gamma \cdot H^2 \cdot Ka \cdot \frac{H}{3}}{\gamma \cdot H \cdot L_r} = \frac{Ka \cdot H^2}{6 \cdot L_r}$$

To meet the condition of a fully compressed base, we have:

$$e = \frac{Ka \cdot H^2}{6 \cdot Lr} \leqslant \frac{L_r}{6} \quad \Rightarrow \quad L_r \geqslant H\sqrt{Ka}$$

$$\Rightarrow \quad L_r \geqslant 5.8 \cdot \tan\left(45° - \frac{35°}{2}\right) = 3.02\,m$$

Thus, the minimum length required for the reinforcements (L_r) to
avoid both sliding and overturning, and still keep the base of the wall
compressed will be 3.02 meters.

e) Load capacity of foundation ground

When considering the formulation of Meyerhof (1955) for the
distribution of acting stresses at the base of the wall, we have:

$$\sigma_{z,b} = \frac{Rz}{L_r - 2e} = \frac{\gamma \cdot H \cdot L_r}{L_r - 2\frac{Ka \cdot H^2}{6 \cdot L_r}} = \frac{\gamma \cdot H}{1 - \frac{Ka}{3} \cdot \left(\frac{H}{L_r}\right)^2}$$

To proceed with the calculations, a length of 4.6 meters (0.8 H)
will be adopted *a priori* for the reinforcements, so as to avoid pullout in

the resistant area. The analysis for pullout resistance will be presented further ahead.

The stress at the base will be:

$$\sigma_{z,b} = \frac{18 \times 5.8}{1 - \frac{0.27}{3} \cdot \left(\frac{5.8}{4.6}\right)^2} = 121.9 \, \text{kN/m}^2$$

Once the safety factor of 2.5 is established for the load capacity of the foundation, the ground must present ultimate stress exceeding 305 kPa. According to Brinch Hansen (1961) and Sokolovski (1960), the load capacity of the ground is given by:

$$q_{ult} = \gamma \cdot D + c \cdot N_c \cdot f_{ci} + \gamma \cdot D \cdot (N_q - 1) \cdot f_{qi} + \frac{1}{2} \cdot \gamma \cdot B' \cdot N_\gamma \cdot f_{\gamma i}$$

where q_{ult} is the ultimate bearing capacity of the foundation soil; D is the embedding of the RSRS foundation; $N_{c,q,\gamma}$ are the bearing capacity factors; $f_{ci,qi,\gamma i}$ are the eccentricity and inclination factors of the load; and B' is the effective width of the foundation $(L_r - 2 \cdot e)$.

The inclination and eccentricity factors are:

$$f_{qi} = \left[1 - \frac{R_h}{R_v + B' \cdot c' \cdot \cotan(\phi)}\right]^2$$

$$f_{ci} = f_{qi} - \frac{1 - f_{qi}}{N_c \cdot \tan(\phi)}$$

$$f_{\gamma i} = (f_{qi})^{3/2}$$

where R_h and R_v are, respectively, the resultants from horizontal and vertical stresses and forces.

Table 5.1 provides values of load capacity factors for strip footings, according to Prandtl (1921), Reissner (1924) and Vesic (1975).

Once the vertical resultant is admitted to be equal to the weight of the reinforced soil mass and the horizontal resultant to be equal to the active thrust, we have:

$$f_{qi} = \left[1 - \frac{82.0}{480.2 + 3.94 \cdot 10 \cdot \cotan(35)}\right]^2 = 0.718$$

$$f_{ci} = 0.718 - \frac{1 - 0.718}{46.1 \cdot \tan(35)} = 0.709$$

$$f_{\gamma i} = (0.718)^{3/2} = 0.608$$

TABLE 5.1 BEARING CAPACITY FACTORS

Effective friction angle (°)	N_c	N_q	N_γ
0	5.14	1.00	0.00
15	10.98	3.94	2.65
20	14.83	6.40	5.39
25	20.72	10.66	10.88
30	30.14	18.40	22.40
35	46.12	33.30	48.03
40	75.31	64.20	109.41
45	133.9	134.9	271.76

Adopting *a priori* a 0.4-meter embedment (equal to the height of two blocks), the bearing capacity of the foundation soil shall be:

$$q_{ult} = 18 \times 0.4 + 10 \times 46.12 \times 0.709 + 18 \times 0.4 \times (33.3 - 1)$$
$$\times 0.718 + \frac{1}{2} \times 18 \times 3.94 \times 48.03 \times 0.608 = 1536\,kPa$$

The safety factor for the load capacity of the foundation may be defined as:

$$FS = \frac{q_{ult}}{\sigma'_{z,b}} = \frac{1536}{121.9} = 12.6$$

The high safety factor that was obtained is a consequence of high parameters adopted for the foundation soil. Given this situation, the embedment is not necessary to increase bearing capacity. However, it is advisable to dig the most superficial layer of the foundation soil to remove organic soils and provide an embedment height equivalent to at least two blocks, in order to prevent future undermining, in the event of inadvertent excavation in front of the toe of the wall.

5.3 INTERNAL STABILITY ANALYSIS

Reinforcements should be designed to avoid reinforcement failure or pullout in the resistant zone (Figure 3.2). The length, the resistance and the amount of reinforcements to be used are determined based on the maximum tensile force acting on the reinforcements, T_{max}.

To determine the stiffness modulus, J_r, of the reinforcement, it is necessary to choose a specific reinforcement. Therefore, one should know T_{max}. Since T_{max} is also a function of J_r, it is necessary to carry out an iterative calculation. Fortunately, the convergence is fast and two or three iterations are sufficient.

The calculations are carried out for the final construction condition, considering the depth corresponding to each layer of reinforcements. The safety factors, FS, were calculated by the conventional procedure (Christopher et al., 1990).

a) Determining the vertical stress induced by compaction

For the compactor rollers used, the vertical stress induced by compaction is 110 kPa (Figure 2.18).

b) Determining the geostatic vertical stress at the reinforcement level

Each reinforcement is located at a generic depth z. According to the formulation of Meyerhof (1955), we have:

$$\sigma'_z = \frac{\gamma' \cdot z}{1 - \left(\frac{K_a}{3}\right) \cdot \left(\frac{z}{L_r}\right)^2} = \frac{18 \cdot z}{1 - 0.0043 \cdot z^2}$$

c) Determining σ'_{zc}

For depths where $\sigma'_z < \sigma'_{zc,i}$, i.e., $\sigma'_z < 110$ kPa, we have: $\sigma'_{zc} = 110$ kPa. For greater depths, where $\sigma'_z > 110$ kPa, we have $\sigma'_{zc} = \sigma'_z$.

d) Determining β

For the first iteration, one should adopt an appropriate S_i, value, according to the type of reinforcement, as listed previously. In this particular case (PVA geogrids), one can consider $S_i = 0.03$.

$$\beta = \frac{\left(\frac{\sigma'_{zc}}{P_a}\right)^n}{S_i}$$

For the first iteration:

$$\beta = \frac{\left(\frac{\sigma'_{zc}}{101.3}\right)^{0.78}}{0.03} = 0.92 \cdot (\sigma'_{zc})^{0.78}$$

For the other iterations:

$$\beta = \frac{\left(\frac{\sigma'_{zc}}{101.3}\right)^{0.78}}{S_i}$$

where $S_i = \dfrac{J_r}{k \cdot P_a \cdot S_v} = \dfrac{J_r}{128 \times 101.3 \times 0.6} = \dfrac{J_r}{7781}$.

e) Determining T_{max}

The maximum tension (T_{max}) is determined for each level of reinforcement, considering the values of β, σ'_z and σ'_{zc} from the charts of Figures A.5, A.6 and A.7 of the Appendix (Dantas and Ehrlich, 1999).

For the first iteration from the arbitrated value, we obtained the values listed in Table 5.2.

TABLE 5.2 SPREADSHEET FOR CALCULATION OF FIRST ITERATION

Reinforce-ment layer	S_v (m)	Elevation (m)	L_r (m)	S_i (kPa)	σ_z (kPa)	σ_{zc}	σ_z/σ_{zc}	β	χ (chart)	T_{max} (kN)
10	0.7	5.40	4.6	0.03	7.2	110.0	0.07	35.5	0.14	11.01
9	0.6	4.80	4.6	0.03	18.1	110.0	0.16	35.5	0.14	9.52
8	0.6	4.20	4.6	0.03	29.1	110.0	0.26	35.5	0.15	9.61
7	0.6	3.60	4.6	0.03	40.4	110.0	0.37	35.5	0.15	9.69
6	0.6	3.00	4.6	0.03	52.1	110.0	0.47	35.5	0.15	9.77
5	0.6	2.40	4.6	0.03	64.4	110.0	0.59	35.5	0.15	9.86
4	0.6	1.80	4.6	0.03	77.3	110.0	0.70	35.5	0.15	9.94
3	0.6	1.20	4.6	0.03	91.0	110.0	0.83	35.5	0.15	10.03
2	0.6	0.60	4.6	0.03	105.8	110.0	0.96	35.5	0.15	10.11
1 (base)	0.3	0.00	4.6	0.03	121.9	121.9	1.00	38.5	0.15	5.65

f) Determining the characteristics of reinforcements

Based on the values of T_{max} obtained in the first iteration, it is possible to determine the appropriate characteristics of the reinforcement considering the safety factor and the calculation resistances for reinforcement failure and pullout, according to items (h) and (i). In this example, we selected the reinforcement with nominal characteristics indicated in Table 5.3.

g) Refinement of the calculation

It is likely that the value of S_i corresponding to the chosen reinforcement will be different from the arbitrated value in the first iteration. Then a second iteration should be carried out, with the new S_i value, resulting in a new value for β and new readings in the charts, which will lead to a new value of T_{max}.

TABLE 5.3 CHARACTERISTICS OF THE CHOSEN REINFORCEMENT (Huesker, 2007)

Feature	Fortrac 35 MP
Tensile strength (kN/m)	35.0
Elongation at burst (%)	4 a 6%
J_r (kN/m)	700
Safety factor for installation damage in sandy, silty or clayey soils	1.05
Safety factor for creep (td = 75 years)	1.514
Safety factor for chemical degradation (2 < pH <13)	1.1
Safety factor for biological degradation	1.0
Project strength/resistance (T_d)	20.0

If this is compatible with the chosen reinforcement, the calculation is finished. Otherwise, a new reinforcement is chosen and one more iteration is carried out.

In this case, only two iterations were necessary. The calculations corresponding to the second iteration are presented in Table 5.4.

TABLE 5.4 SPREADSHEET FOR CALCULATION OF SECOND ITERATION

Reinforcement layer	S_v (m)	Elevation (m)	J_r (kN/m)	L_r (m)	S_i	σ_z (kPa)	σ_{zc} (kPa)	σ_z/σ_{zc}	β	χ (chart)	T_{max} (kN)
10	0.7	5.40	700	4.6	0.077	7.2	110.0	0.07	13.8	0.15	11.70
9	0.6	4.80	700	4.6	0.090	18.1	110.0	0.16	11.9	0.16	10.41
8	0.6	4.20	700	4.6	0.090	29.1	110.0	0.26	11.9	0.16	10.69
7	0.6	3.60	700	4.6	0.090	40.4	110.0	0.37	11.9	0.17	11.07
6	0.6	3.00	700	4.6	0.090	52.1	110.0	0.47	11.9	0.17	11.46
5	0.6	2.40	700	4.6	0.090	64.4	110.0	0.59	11.9	0.18	11.84
4	0.6	1.80	700	4.6	0.090	77.3	110.0	0.70	11.9	0.19	12.22
3	0.6	1.20	700	4.6	0.090	91.0	110.0	0.83	11.9	0.19	12.60
2	0.6	0.60	700	4.6	0.090	105.8	110.0	0.96	11.9	0.20	13.17
1 (base)	0.3	0.00	700	4.6	0.180	121.9	121.9	1.00	6.4	0.22	7.93

h) Allowable strength of reinforcement

Once a minimum safety factor of 1.5 is established for reinforcement failures, geogrids with resistance calculation (T_d) of at least 13.17 kN/m × 1.5 = 19.8 kN/m will be made necessary. Therefore, the chosen reinforcement can be used, as it has T_d = 20.0 kN/m.

i) Pullout stability

First, it is necessary to determine the length of the available embedment beyond the active wedge (L_e), which depends on the depth of each reinforcement:

$$L_e = L_r - (H - z) \cdot \left[\tan \left(45° - \frac{\phi}{2} \right) - \frac{1}{\tan \omega} \right] = 4.6 - 0.42 \, (5.8 - z)$$

Pullout resistance (P_r) can be determined based on the characteristics of the soil-reinforcement contact and geostatic stresses acting on each vertical reinforcement:

$$P_r = 2F^* \cdot \alpha \cdot \sigma_v' \cdot L_e \quad \geqslant \quad FS \cdot T_{max}$$

The scale factor (α) will be allowed as 1.00 and the pullout resistance factor (F^*) can be determined by:

$$F^* = f_a \cdot \tan \phi = 0.8 \cdot \tan 35° = 0.56$$

Thus, we have:

$$P_r = 2 \cdot 0.56 \cdot 1.0 \cdot \sigma_z' \cdot [4.6 - 0.42 \cdot (5.8 - z)]$$

$$FS = \frac{P_r}{T_{max}}$$

Table 5.5 presents a synthesis of safety factors for pullout. The 1.5 minimum safety factor was met at all reinforcement levels.

TABLE 5.5 PULLOUT SAFETY FACTORS

Reinforcement layer	depth (m)	L_e (m)	P_r (kN/m)	T_{max} (kN/m)	FS
10	0.40	2.33	18.80	11.70	1.6
9	1.00	2.58	52.28	10.41	5.0
8	1.60	2.83	92.44	10.69	8.6
7	2.20	3.09	139.80	11.07	12.6
6	2.80	3.34	195.02	11.46	17.0
5	3.40	3.59	258.97	11.84	21.9
4	4.00	3.84	332.71	12.22	27.2
3	4.60	4.10	417.62	12.60	33.2
2	5.20	4.35	515.40	13.17	39.1
1 (base)	5.80	4.60	628.25	7.93	79.2

5.4 COMPARISON OF METHODS FOR REINFORCEMENT TENSION DETERMINATION

It is interesting to compare the results above with the results presented by other methods. In Figure 5.2, a comparison between the T_{max} values obtained above and the values that would be obtained by conventional design methods based on limit equilibrium or semi-empirical considerations is presented. Cohesion was dismissed because the two limit equilibrium methods did not take it into account. As is traditional when using these methods, compaction efforts were ignored.

It is noted that the methods based on limit equilibrium or semi-empirical considerations tend to present lower tensile strength values on the upper reinforcements and higher values on the lower reinforcements, for ignoring the influence of soil compaction.

Whenever high compaction energy is used, this effect will be against safety, because it may induce the designer to specify reinforcements with lesser resistance than required in the upper layers. It is worth noting that the estimation of tensile strength near zero in the uppermost reinforcement level goes against practical experience.

Figure 5.3 presents a similar comparison, but considers compaction as an evenly distributed surcharge. In this case, it is noted that the conventional analyzed methods showed much higher T_{max} values

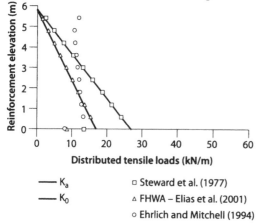

— K_a □ Steward et al. (1977)
— K_0 △ FHWA – Elias et al. (2001)
                   ○ Ehrlich and Mitchell (1994)

Figure 5.2 *Comparison between T_{max} values obtained through various methods, ignoring compaction in limit equilibrium methods*

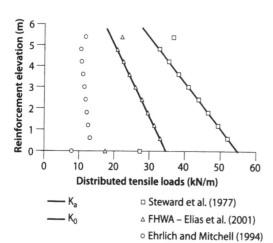

— K_a □ Steward et al. (1977)
— K_0 △ FHWA – Elias et al. (2001)
                   ○ Ehrlich and Mitchell (1994)

Figure 5.3 *Comparison between T_{max} values obtained through various methods, using the technique of considering compaction as distributed surcharge in limit equilibrium methods*

than those obtained through the method of Dantas and Ehrlich (1999).

The technique of considering the compaction as distributed surcharge when using conventional methods that were not developed for this situation led to the estimation of exaggerated tensile strengths.

5.5 WALL FINAL ARRANGEMENT

The final arrangement of the retaining structure can be seen in Figure 5.4.

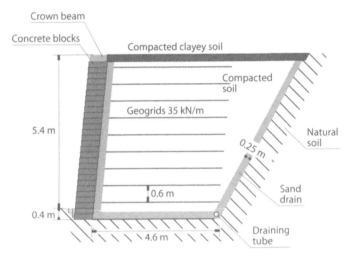

Figure 5.4 *Typical cross section of the designed wall*

Appendix – Design charts

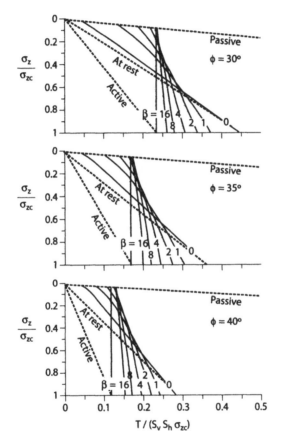

Figure A.1 *Charts for determining "χ" for the calculation of T$_{max}$ in structures with vertical facing (Ehrlich and Mitchell, 1994)*

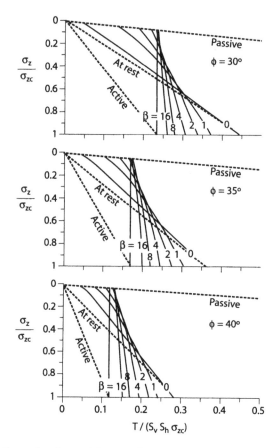

Figure A.2 *Charts for determining "χ" for the calculation of T_{max} in structures with face inclination 3V:1H (Dantas and Ehrlich, 2000a)*

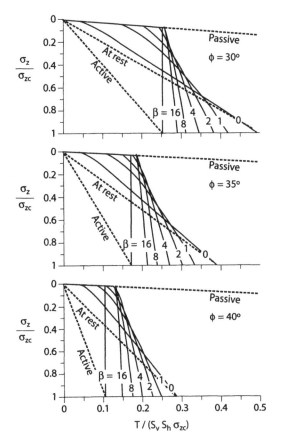

Figure A.3 *Charts for determining "χ" for the calculation of* T_{max} *in structures with face inclination 2V:1H (Dantas and Ehrlich, 2000a)*

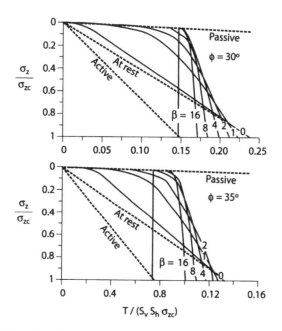

Figure A.4 *Charts for determining "χ" for the calculation of T$_{max}$ in structures with face inclination 1V:1H (Dantas and Ehrlich, 2000a)*

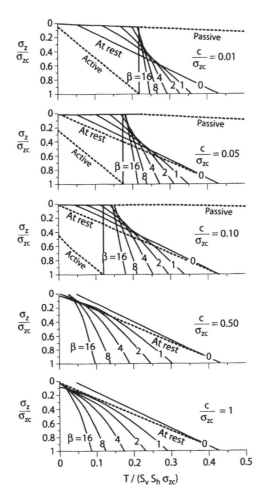

Figure A.5 *Charts for determining* T_{max}, *considering soil cohesion (Dantas and Ehrlich, 1999) for structures with vertical face inclinations, with* $\phi' = 35°$

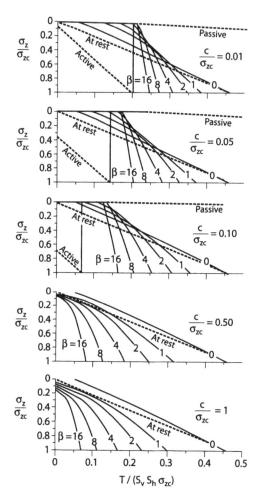

Figure A.6 *Charts for determining* T_{max}, *considering soil cohesion (Dantas and Ehrlich, 1999) for structures with 3V:1H face inclinations, with* $\phi' = 35°$

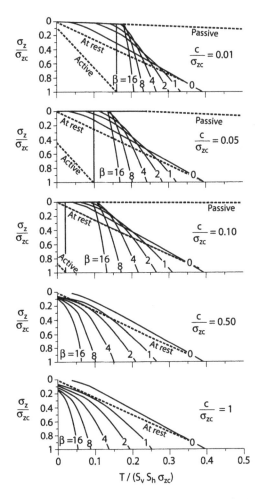

Figure A.7 *Charts for determining T_{max}, considering soil cohesion (Dantas and Ehrlich, 1999) for structures with 2V:1H vertical face inclinations, with $\phi' = 35°$*

References

ABNT - Brazilian Technical Standards Association. Soil - Determination of liquid limit. Rio de Janeiro: ABNT, 1984.

ABNT - Brazilian Technical Standards Association. Soil grains finer than 4.8mm - Determination of density. Rio de Janeiro: ABNT, 1984.

ABNT - Brazilian Technical Standards Association. Soil - Determination of plastic limit. Rio de Janeiro: ABNT, 1984.

ABNT - Brazilian Technical Standards Association. Soil - Grain size analysis. Rio de Janeiro: ABNT, 1984.

ABNT - Brazilian Technical Standards Association. Soil - Compaction test. Rio de Janeiro: ABNT, 1986.

ABNT - Brazilian Technical Standards Association. Soil Samples - Preparation for compaction and characterization tests. Rio de Janeiro: ABNT, 1986.

ABNT - Brazilian Technical Standards Association. Terre Armée. Rio de Janeiro: ABNT, 1986.

ABNT - Brazilian Technical Standards Association. Determination of the wide width tensile resistance - Method of test. Rio de Janeiro: ABNT, 1993.

ABNT - Brazilian Technical Standards Association. Geotextiles - Testing for determination of static puncture strenght - Method of test. Rio de Janeiro: ABNT, 1995.

ABNT - Brazilian Technical Standards Association. Soil - Dispersive characteristics of clay soil by double hydrometer - Method of test. Rio de Janeiro: ABNT, 1996.

ABNT - Brazilian Technical Standards Association. Geosynthetics - Determination of tensile creep and creep rupture behaviour. Rio de Janeiro: ABNT, 2005.

ABRAMENTO, M.; WHITTLE, A. J. Shear-lag analysis of a planar soil reinforcement in plane strain compression. *Journal of Engineering Mechanics*, ASCE, v. 119, n. 2, p. 270-291, 1993.

ADIB, M. E. *Internal lateral earth pressure in earth walls.* Civil Engineering Dept., University of California, Berkeley, 1988.

ALEXIEW, D.; SILVA, A. E. F. Discussion on limit equilibrium analysis models and modes for reinforced soil slopes and walls. In: Brazilian Symposium on Geosynthetics, 4., 2003, Porto Alegre. *Proceedings...* Porto Alegre, 2003.

ALLEN, T. M. Determination of long-term tensile strength of geosynthetics: a state of the art rewiew. In: GEOSYNTHETICS'91 CONFERENCE, 1991, Atlanta. *Proceedings...* Atlanta: IFAI Publication, 1991. p. 351-380.

ANDRADE, P. I.; EHRLICH, M.; ITURRI, E. A. Z. Estudo numérico da influência de carregamentos externos nas tensões atuantes em murros de solo reforçado. In: 1st South-American Symposium on Geosynthetics; 3rd Brazilian Symposium on Geosynthetics, 1999, Rio de Janeiro. *Proceedings...* Rio de Janeiro, 1999.

AUTRAMAQ. *New VAP70 compactor roller.* (in Portuguese) 8 Sep. 2007a <http://www.autramaq.com.br/rolo_compactador_vap70.htm>.

AUTRAMAQ. *VAP55 compactor roller.* (in Portuguese) 8 Sep. 2007b <http://www.autramaq.com.br/rolo_compactador_vap55.htm>.

AZAMBUJA, E. A influência do dano mecânico na determinação na tensão admissível dos geossintéticos em estruturas de solo reforçado. In: 1st South-American Symposium on Geosynthetics; 3rd Brazilian Symposium on Geosynthetics, 1999, Rio de Janeiro. *Proceedings...* Rio de Janeiro, 1999. p. 157-165.

AZAMBUJA, E.; STRAUSS, M. Casos históricos de sistemas de contenção em solo reforçado com geossintéticos no Rio Grande do Sul - Brasil. In: 1st South-American Symposium on Geosynthetics; 3rd Brazilian Symposium on Geosynthetics, 1999, Rio de Janeiro. *Proceedings...* Rio de Janeiro, 1999. p. 387-394.

BECKER, L. D. B. *Creep of geotextiles confined in experimental fill.* 2001. 122f. MSc. Thesis (In Portuguese). Federal University of Rio Grande do Sul, Porto Alegre, 2001.

BECKER, L. D. B. *Behavior of geogrids in a reinforced wall and in pullout tests.* 2006. 322 p. D.Sc. Thesis. (in Portuguese). PUC-Rio, Pontifical Catholic University of Rio de Janeiro, Brazil, 322p.

BENJAMIN, C. V. S. *Experimental study of geotextile reinforced soil structures.* 2006. 326 p. D.Sc. Thesis. (in Portuguese). University of São Paulo, São Carlos, 326p.

BERGADO, D. T.; CHAI, J. C. Pullout force/displacement relationship of extensible grid reinforcements. *Geotextiles and Geomembranes Journal,* Amsterdam, v. 13, p. 295-316, 1994.

BERGADO, D. T. et al. Interaction between cohesive-frictional soil and various grid reinforcements. *Geotextiles and Geomembranes Journal,* Amsterdam, v. 12, p. 327-349, 1993.

BRINCH HANSEN, J. The ultimate resistance of rigid piles against transverse forces. Dansk Geotechnisk Inst. Bull, 1961. v. 12, p. 5-9. In: CLAYTON, C. R. I.; MILITITSKY, J.; WOODS, R. I. *Earth pressure and Earth-retaining structures.* 2. ed. Glasgow: Blackie Academic and Professional, 1993.

BRUGGER, P. J. *Personal archive,* 2007.

BRUNO, A. C.; EHRLICH, M. Performance of a geotextile reinforced soil wall. In: PAN-AMERICAN SYMPOSIUM ON LANDSLIDES, 2., 1997, Rio de Janeiro. *Proceedings...* Rio de Janeiro,1997. p. 665-670.

CARVALHO, P. A.; PEDROSA, J. A. B. A.; WOLLE, C. M. Fill reinforcement with geotextiles - an alternative for geotechnical engineering. (In Portuguese) In: Brazilian Conference of Soil Mechanics and Foundation Engineering, 1986, Porto Alegre. Proceedings... Porto Alegre, 1986. v.4 p. 169-178.

CASE CONSTRUCTION. Technical specifications catalogue. 8 Sep. 2007 <http://www.casece.com/wps/portal/casece?brandsite_-brand=CaseCEebrandsite_language=enebrandsite_geo=NA>.

CAZZUFFI, D.; GHINELLI, A.; SACCHETTI, M.; VILLA, C. European experimental approach to the tensile creep behaviour of high strengh geosynthetics. In: GEOSYNTHETIC'97 CONFERENCE, 1997, Long Beach, California. *Proceedings...* 1997. p. 253-266.

CHRISTOPHER, B. R.; GILL, S. A.; GIROUD, J. P.; JURAN, I.; MITCHELL, J. K.; SCHLOSSER, F.; DUNNICLIFF, J. *Reinforced soil structures,* v. 1 - Design and construction guidelines. Washington, D.C.: FHWA, 1990. Rep. n. FHWA/RD/89-043.

CLAYTON, C. R. I.; MILITITSKY, J.; WOODS, R. I. *Earth pressure and Earth-retaining structures.* 2. ed. Glasgow: Blackie Academic and Professional, 1993.

COZZOLINO, V. M. N.; NOGAMI, J. S. MCT geotechnical classification for tropical soils. Soils and Rocks (In Portuguese), v. 16, pp. 77-91, 1993.

DANTAS, B. T. *Working stress analysis for reinforced soil slopes.* 1998. MSc. Thesis (In Portuguese). COPPE/Federal University of Rio de Janeiro, Rio de Janeiro, 1998.

DANTAS, B. T. *Analysis of the Behavior of Reinforced Soil Structures Under Working Stress Conditions.* 2004. 222p. D.Sc. Thesis. (in Portuguese). COPPE/Federal University of Rio de Janeiro, Rio de Janeiro, 2004.

DANTAS, B. T.; EHRLICH, M. Charts for working stress design of reinforced slopes (In Portuguese).

DANTAS, B. T.; EHRLICH, M. ANALYSIS method for reinforced slopes UNDER WORKING STRESS CONDITIONS. Soils and Rocks (In Portuguese), Rio de Janeiro, v.23, n.2, pp. 113-133, 2000a.

DANTAS, B. T.; EHRLICH, M. Perfomance of geosynthetic reinforced slopes at failure. *J.Geot. e Geoenv. Engrg.* ASCE, Reston, Virginia, v. 126, n. 3, p. 286-288, mar. 2000b.

DEN HOEDT, G. Creep and relaxation of geotextile fabrics. *Geotextiles and Geomembranes Journal,* Amsterdam, v. 4, n. 2, p. 83-92, 1986.

Brazilian Department for Transportation infrastructure. Brazilian Standard DNER-ME 28/61: quick stove-top method for determining the water content of soils (In Portuguese). 1961.

Brazilian Department for Transportation infrastructure. Brazilian Standard DNER-ME 88/64: Quick alcohol method for determining water content of soils (In Portuguese). 1964.

Brazilian Department for Transportation infrastructure. Brazilian Standard DNER-ME 52/94: Soils and small aggregates - Determination of water content through the use of "Speedy" (In Portuguese). 1994.

DUNCAN, J. M.; SEED, R. B. Compaction-induced earth pressures under K_0-conditions. *Journal of Geotechnical Engineering,* ASCE, Reston, Virginia, v. 112, n. 1, p. 1-22, 1986.

DUNCAN, J. M. et al. *Strength, stress-strain and bulk modulus parameters for finite element analyses of stresses and movements in soil masses. Journal of Geotechnical Engineering,* Rep. No. UCB/GT/80-01, University of California, Berkeley, California, 1980.

DYER, N. R.; MILLIGAN, G. W. E. A photoelastic investigation of the interaction of a cohesionless soil with reinforcement placed at different orientations. In: INT. CONF. ON IN SITU SOIL AND ROCK REINFORCEMENT, 1984, Paris. *Proceedings...* Paris, 1984. p. 257-262.

DYNAPAC. Technical specifications catalogue. 8 Sep. 2007. <http://www.dynapac.com/en/Products/?cat=10eproduct=2>

EHRLICH, M. Deformation of reinforced soil walls. In:2nd Brazilian Symposium on Geosynthetics (In Portuguese). 1995. São Paulo. *Proceedings...* São Paulo: ABMS, 1995. pp. 31-40.

EHRLICH, M. Analysis of reinforced walls and slopes. In: 1st South American Symposium on Geosynthetics, 1999, Rio de Janeiro. *Proceedings...* Keynote Lecture, Rio de Janeiro, 1999. p. 73-84

EHRLICH, M.; AZAMBUJA, E. Reinforced Soil Walls. In:4th Brazilian Symposium on Geosynthetics (In Portuguese). 2003. Porto Alegre. *Proceedings...* Porto Alegre: ABMS, 2003. pp. 81-100.

EHRLICH, M.; MITCHELL, J. K. Working stress design method for reinforced soil walls. *Journal of Geotechniacl Engineering*, ASCE, Reston, Virginia, v. 120, n. 4, p. 625-645, 1994.

EHRLICH, M.; VIANNA, A. J. D.; FUSARO, F. Behavior of a Reinforced Soil Wall. In: 10th Brazilian Conference of Soil Mechanics and Foundation Engineering (In Portuguese), 1994, Foz do Iguaçu. *Proceedings...* pp. 819-824.

ELIAS, V.; CHRISTOPHER, B. R.; BERG, R. R. Mechanically stabilized earth walls and reinforced soil slopes - design and construction guidelines. *Geotechnical Engineering*, Washington, n. FHWA-NHI-00-043, p. 394, 2001.

HUESKER. Technical specifications catalogue. (in Portuguese) 8 Sep. 2007 <http://www.usconstructionfabrics.com/huesker.aspx>.

JANBU, N. Soil compressibility as determined by oedometer and triaxial tests. In: EUROPEAN CONFERENCE ON SOIL MECHANICS AND FOUNDATION ENGINEERING, 1963, Wiesbaden. *Proceedings...* Wiesbaden, Germany, 1963. v. 1, p. 19-25; v. 2, p. 17-21.

JEWELL, R. A. *Some effects of reinforcement on the mechanical behavior of soils.* 1980. Ph.D. dissertation – University of Cambridge, Cambridge, England, 1980.

JEWELL, R. A. Application of revised design charts for steep reinforced slopes. *Geotextiles and Geomembranes Journal*, v. 10, p. 203-233, 1991.

JEWELL, R. A. et al. Interaction between soil and geogrids. In: SYMPOSIUM ON POLYMER GRID REINFORCEMENT IN CIVIL ENGINEERING, 1984, United Kingdom. *Proceedings...* United Kingdom, 1984. p. 18-30.

JURAN, I.; CHEN, C. L. Soil-geotextile pullout interaction properties: testing and interpretation. *Transportation Research Record*, Washington, n. 1188, p. 37-47, 1988.

KOERNER, R. M. *Designing with geosynthetics*. 2. ed. New Jersey: Prentice-Hall, 1990.

KOERNER, R. M. *Designing with geosynthetics*. 4. ed. New Jersey: Prentice-Hall, 1998.

KONDNER R. L.; ZELASKO J. S. A hyperbolic stress-strain formulation of sands. In: PAN-AMERICAN CONFERENCE ON SOIL MECHANICS AND FOUNDATION ENGINEERING, 2., 1963, São Paulo. *Proceedings...* São Paulo, 1963. p. 289-324.

LESHCHINSKY, D.; BOEDEKER, R.H. Geosynthetic reinforced soil structures. *Journal of Geotechniacl Engineering*, ASCE, Reston, Virginia, v. 115, n.10, p. 1459-1478, 1989.

MAIOLINO, A. L. G. *Shear resistance of campacted soils: a proposal for classification*. 1985. MSc. Thesis (In Portuguese). COPPE/Federal University of Rio de Janeiro, Rio de Janeiro, 1985.

MARQUES, H. C.; EHRLICH, M.; RICCIO, M. V. Shear resistance and deformability parameters for finite element analysis of compacted tropical soils from Brazil. Internal report (In Portuguese). Rio de Janeiro: COPPE/Federal University of Rio de Janeiro. 2006.

MCGOWN, A.; ANDRAWES, K. Z.; KABIR M. H. Load extension testing of geotextiles confined in soil. In: INTERNATIONAL CONFERENCE ON GEOTEXTILES, 2., 1982, Los Angeles. *Proceedings...* Los Angeles, 1982. v. 3. p. 793-798.

MELO, L. V. *Comportamento Tensão-deformação-resistência de solos compactados dos núcleos das barragens Eng. Armando Ribeiro Gonçalves (RN) e Bocaina (PI)*. 1986. Dissertação (Mestrado) – COPPE/UFRJ, Rio de Janeiro, 1986.

MEYERHOF, G. G. The bearing capacity of foundations under eccentric and inclined loads. In: INT. CONF. ON SOIL MECH. AND FOUND. ENGIN., 3., 1955, Zurich. *Proceedings...* Zurich, Switzerland, 1955. v. 1. p. 440-445.

MILLIGAN, G. W. E.; PALMEIRA, E. M. Prediction of bond between soil and reinforcement. In: INTERNATIONAL SYMPOSIUM ON PREDICTION AND PERFORMANCE ON GEOTEXTILE ENGINEERING, 1987, Calgary. *Proceedings...* Calgary, Canada, 1987. p. 147-153.

MITCHELL, J. K.; VILLET, W. C. B. Reinforcement of earth slopes and embankments. *Transportation Research Board*, NCHRP, Report 290, Washington, 1987.

MÜLLER-ROCHHOLZ, J.; REINHARD, K. Creep of geotextiles at different temperatures. In: INTERNATIONAL CONFERENCE ON GEOTEXTILES, GEOMEMBRANES AND RELATED PRODUCTS, 4., 1990, Hague. *Proceedings...* Hague, Netherlands, 1990. p. 657-659.

NOGAMI, J. S.; VILLIBOR, D. F. A new classification of soils for highway purposes. In: Brazilian Symposium on engineering use of tropical soils (In Portuguese). 1981. Rio de Janeiro. *Proceedings...* Rio de Janeiro: COPPE/UFRJ. 1981. v. 1. pp. 30-41.

NOGAMI, J. S.; VILLIBOR, D. F. *Low cost pavement with lateritic soils* (In Portuguese). São Paulo: Villibor. v. 1. 197 p.

PALMEIRA, E. M.; GOMES, R. C. Comparisons of predicted and observed failure mechanisms in model reinforced soil walls. *Geosynthetics International*, v. 3, n. 3, p. 329-347, 1996.

PERALTA, F. N. G. Comparison of design methods for geosynthetics reinforced soil wall. Rio de Janeiro, 2007. 162p. M.Sc. Dissertation (In Portuguese) - Civil Engineering Departament, Pontifical Catholic University of Rio de Janeiro.

PRANDTL, L. Über die eindrigungsestigkeitplasticher baustoffe und die festigkeit von shneiden. Zeitsch. *Angew Mathematik und Mechanik*. 1921. p. 15-20. In: CLAYTON, C. R. I.; MILITITSKY, J.; WOODS, R. I. *Earth pressure and Earth-retaining structures.* 2. ed. Glasgow: Blackie Academic and Professional, 1993.

REISSNER, H. Zum Erddruckproblem. In: INTERNATIONAL CONFERENCE ON APPLIED MECHANICS, 1., 1924, Delft. *Proceedings...* Delft, 1924. p. 295-311. In: CLAYTON, C. R. I.; MILITITSKY, J.; WOODS, R. I. *Earth pressure and Earth-retaining structures.* 2. ed. Glasgow: Blackie Academic and Professional, 1993.

RICCIO, F. M. V. *Behavior of Reinforced Soil Wall With Fine-Grained Tropical Soils.* 2007. 441 p. D.Sc. Thesis (In Portuguese) COPPE/Federal University of Rio de Janeiro. Rio de Janeiro, 2007.

SAYÃO, A. S. F. J. Reinforced Walls and Slopes. In: VERTEMATTI, J. C. Brazilian Manual of geosyntheticson Manual brasileiro de geossint (In Portuguese). São Paulo: Edgard Blücher, 2004. pp. 84-123.

SEED, R. B.; DUNCAN, J. M. FE Analyses: compaction-induced stresses and deformations. *Journal of Geotechnical Engineering*, ASCE, v. 122, n. 1, p. 23-43, 1986.

SILVA, A. E. F.; VIDAL, D. M. Reinforced Soil Structures and limit equilibrium design methods In: SOUTH AMERICAN SYMPOSIUM ON GEOSYNTHETICS, 1., 1999, Rio de Janeiro. *Proceedings...* Rio de Janeiro, 1999. pp. 139-147.

SOKOLOVSKI, V. V. *Statics of soil media*. London: Butterworth, 1960. In: CLAYTON, C. R. I.; MILITITSKY, J.; WOODS, R. I. *Earth pressure and Earth-retaining structures*. 2. ed. Glasgow: Blackie Academic and Professional, 1993.

STEWARD, J. E.; WILLIANSON, R.; MOHNEY, J. Earth Reinforcement, chapter 5. *Guidelines for use of fabrics in construction and maintenace of low* – volume Roads. Portland, Oregon: USDA Forest Service, 1977.

TEIXEIRA, S. H. C. *Study of soil-geogrid interaction in pullout tests and its use on design and analysis of reinforced soil structures*. 2003. 214p. D.Sc. Thesis. (in Portuguese). University of São Paulo, São Paulo, 2003.

TERZAGHI, K.; PECK, R. B. *Soil mechanics in engineering practice*. 2. ed. New York: John Wiley & Sons, 1967.

VESIC, A. S. Bearing capacity of shallow foundations. In: *Foundation Engineering Handbook*. New York: McGraw-Hill, 1975.

VIDAL, H. The principle of reinforced earth. *Highway Research Record*, n. 282, 1969.

WACKER Construction Equipment AG. Technical specifications catalogue. 8 Sep. 2007 <http://products.wackerneuson.com/webapp/ecomm/products/ProdGuide.jsp?source=machines>.

WATTS, G. R. A.; BRADY K. C.; GREENE, M. J. The creep of geosynthetics – prepared for civil engineering, highways agency. *Transport Research Laboratory Report*, v. 319, p. 40, 1998.

WILSON-FAHMY R. F.; KOERNER, R. M. Finite element modelling of soil-geogrid interaction with application to the behaviour of geogrids in a pull-out loading condition. *Geotextiles and Geomembranes Journal*, v. 12, p. 479-501, 1993.

ZORNBERG, J. SITTAR; N.; MITCHELL, J. K. Performance of geosynthetic reinforced slopes at failure. *J. Geot. and Geoenv. Engin.*, ASCE, v. 124, n. 8, p. 670-683, 1999.

⊞ HUESKER

Skillful design, successful delivery

HUESKER GEOSYNTHETICS

Geogrids, wovens, geosynthetic clay liners, nonwovens, drainage and other composites as well as erosion protection products to meet all your needs.

HUESKER engineers can assist in the design of the most demanding projects. Extensive knowledge with many years' experience provide the basis for a successful construction project delivery.

Consult HUESKER for products and engineering solutions!

HUESKER geosynthetics – reliability through experience

www.huesker.com

SOIL REINFORCEMENT

ROAD AND RAILWAY CONSTRUCTION

HYDRAULIC ENGINEERING

ENVIRONMENTAL TECHNOLOGY

HUESKER Synthetic GmbH
48712 Gescher | Germany

Tel.: + 49 (0) 25 42 / 701 - 0
info@huesker.de

HUESKER Synthetic GmbH
www.huesker.com

HUESKER designs, manufactures and markets custom-made geosynthetics for the construction industry worldwide – economical, safe and progressive.

In over 40 years of close cooperation with its customers, consultant engineers, research institutions and testing laboratories, HUESKER has developed individual solutions for various engineering applications in addition to the standard product range.

These applications range from - but are not limited to - soil reinforced foundation construction, the economic construction of transportation infrastructure and hydraulic structures. The task of developing environmentally safe, but economical, landfills and the improvements of roads using asphalt reinforcement, are further examples of HUESKER's diverse activities

The product portfolio offered by HUESKER is as diverse as its applications:

Fortrac®	flexible, high-modulus and low-creep geogrid for soil reinforcement
Fortrac 3D®	flexible, three-dimensional reinforcement grid with erosion protection
HaTelit®	flexible, high-modulus and temperature resistant grid for asphalt reinforcement
Stabilenka®	high-modulus polyester woven for reinforcement and separation of soils
Robutec®	very high-modulus and alkaliresistant woven for reinforcement and separation of soils
Fornit®	biaxial geogrid for subbase reinforcement
Comtrac®	geocomposite for reinforcement, separation and filtration of soils
Duogrid®	geocomposite made of biaxial, high-modulus, flexible geogrid and a nonwoven
NaBento®	geosynthetic clay liner (GCL) for sealing
Incomat®	concrete- or sandmat for sealing and erosion control
Ringtrac®	Geotextile tube for reinforcement and soil containment
HaTe®	Wovens and nonwovens for separation, filtration, drainage and protection
SoilTain®	Systems for hydraulic engineering and dewatering